Ritual as a Missing Link

Ritual as a Missing Link

Sociology, Structural Ritualization Theory and Research

J. David Knottnerus

Routledge
Taylor & Francis Group
LONDON AND NEW YORK

First published 2012 by Paradigm Publishers

Published 2016 by Routledge
2 Park Square, Milton Park, Abingdon, Oxon OX14 4RN
711 Third Avenue, New York, NY 10017, USA

Routledge is an imprint of the Taylor & Francis Group, an informa business

Library of Congress Cataloging-in-Publication Data

Knottnerus, J. David.
Ritual as a missing link : sociology, structural ritualization theory, and research / J. David Knottnerus.
p. cm.
Includes bibliographical references and index.
ISBN 978-1-59451-856-0 (paperback : alk. paper)
1. Social psychology. 2. Ritualization. 3. Ritual—Social aspects. 4. Social structure. I. Title.
HM1033.K65 2010
302'.14—dc22

2010016327

ISBN 13 : 978-1-59451-855-3 (hbk)
ISBN 13 : 978-1-59451-856-0 (pbk)

Designed and Typeset by Straight Creek Bookmakers.

To my mother

Contents

Preface

Why do people behave the way they do? Sociology provides a number of answers to this question. Some approaches focusing on the micro dimensions of social life emphasize factors such as rational choice, symbolic meanings, emotions, or expectations in their explanations. Other perspectives focus on larger, macro-level phenomena such as organizational, occupational, systemic, stratification, or demographic variables to explain social developments and sometimes social behavior. Approaches of the latter kind have dominated much of the discourse and research within sociology over the last century.

One particular type of concept, however, that has not been fully exploited to explain social phenomena involves the idea of ritual. While our appreciation and understanding of ritual has been greatly enhanced by the seminal contributions of different scholars through the years, such as Durkheim or Goffman, it still remains seriously underutilized for analyzing and studying human behavior and social dynamics. The approach presented in this book, structural ritualization theory, attempts to narrow this gap in our understanding of how we behave and the social causes and consequences of our actions.

This approach focuses on the ritualized behaviors that define much of our daily lives and that are found in different kinds of social situations. The study of such rituals employs the scientific method. More precisely, I take a "broad" approach to science in sociology, which, among other things, recognizes the central role of abstract analytical concepts, the crucial significance of empirical evidence, the value of quite different research methods, and the usefulness of integrating concepts from different theoretical approaches and research traditions. Such an approach is also committed to using this knowledge derived from research to more effectively address problems in modern society. These points have been emphasized by various individuals, especially contributors to the Sociological Imagination Group and the book series Advancing the Sociological Imagination, which are inspired by a number of the arguments of C. Wright Mills.

Stated somewhat differently, structural ritualization theory is grounded in a commitment to three essential goals: the development of theory and abstract concepts, substantiating and illustrating these concepts through an examination of empirical evidence, and the application of this knowledge to social problems and undesirable, dehumanizing conditions in contemporary society.

This is why ritual provides a missing link in sociology. The concept of ritual and the perspective described here deal with these and other issues in the attempt to better explain some of the extreme complexity of human action and social reality—a complexity, I would add, that is often missed due to the specialized nature of sociological research and the numerous sections of the American Sociological Association, which have limited communication among one another. In numerous ways, which are enumerated and discussed in this book, ritual provides a link for sociological analysis and research and expands our understanding of how we are social creatures and why we behave the way we do.

This book, therefore, has several objectives because it represents the first attempt to comprehensively describe structural ritualization theory, a perspective that since its inception about a decade ago has developed in several directions involving different extensions or lines of research. Many of its core concepts, definitions, and principles are presented here. These concepts deal with, for example, factors that influence the dominance and impact of ritualized practices on individuals and groups and a model of levels of the social order that helps us understand how rituals may spread across different micro and more macro dimensions of society. Several lines of research, especially work focusing on how structural reproduction has occurred throughout history and around the world, how rituals can be strategically used and power can influence rituals in different social settings, and how the disruption of ritualized practices and the reconstitution of ritual subsequent to such events are of crucial importance for human beings, are discussed. Other current or recent lines of research using this theory are also identified and briefly described.

Of course, weaving its way through the book and highlighted especially in its beginning and latter parts is a discussion of how this work provides a missing link in sociology. In the last chapter some of the implications of this research for addressing pressing social problems in contemporary society, the development of social policy and interventions, and enriching our personal lives are also addressed. Finally, throughout the book possible future research topics and extensions are suggested. The emphasis on future research is no accident because my firm hope is that work using this approach will only expand in terms of both building on previous lines of research and developing new ideas and topics for study that would involve a wider circle of collaborators and individuals who are interested in thinking about these issues.

In closing I would like to acknowledge the crucial contributions and help of a number of persons in the preparation of this work. Bernard Phillips, who spearheaded the development of the Sociological Imagination Group, has provided me with numerous invaluable suggestions and insights while always strongly encouraging my writing of this book. His support has been critical to its development. I would also like to acknowledge Dean Birkenkamp, Paradigm's publisher, who has been both extremely supportive of and patient with me as I prepared this manuscript.

I most certainly want to acknowledge and express my gratitude to all those I have worked with on research projects, in addition to those who I have discussed this work with, including my students. While I will not name all of these individuals and other interested parties here, those persons I have collaborated with and whose research is discussed in this book are clearly cited. Collaborators include current and former graduate and undergraduate students and colleagues at my own university and other academic institutions. Finally, I would like to express my gratitude to one particular person, my wife, Frédérique Van de Poel-Knottnerus (professor of French and French literature) who collaborated with me on the first study utilizing the theory along with more recent research. She has greatly contributed to my thinking about ritual and has always provided great encouragement to me on all my work dealing with this topic. To all these people and all those scholars who over the last century or more have contributed to our understanding of ritual I am truly indebted.

J. David Knottnerus
Department of Sociology
Oklahoma State University

Chapter 1

Understanding Rituals in Everyday Life

Sociology and Structural Ritualization Theory

The basic assumption that rituals are crucial to human behavior is consistent with the arguments of different social scientists who have used this concept to analyze and examine society such as Durkheim (1965), Goffman (1967), Collins (1981; 1987; 2004), Shils and Young (1953), Bell (1992), Douglas (1970), Rappaport (1999), Kertzer (1988), Victor W. Turner (1967), Hocart (1968), Cohen (1974), Alexander, Giesen, and Mast (2006), Warner (1959; 1962), and Bellah (2005). However, despite the work of these analysts—especially scholars such as Durkheim in his classic work *The Elementary Forms of the Religious Life*—the concept of ritual has been underutilized, if not often ignored, in sociology.

This is due to the conventional understanding of rituals in sociology (and in other social sciences and society at large). For instance, it is often assumed that rituals are found only or far more often in premodern societies, not modern societies. This idea is very likely due in part to implicit or explicit evolutionary assumptions that depict modern societies as increasingly rational, more advanced, more civilized, and not influenced by the traditions and customs associated with the premodern world. It is also often assumed that rituals occur only or mainly in religious or sacred contexts. The religious arena is oftentimes seen as a special domain of social life that is dominated by ritual. So too, rituals are commonly depicted as static, unchanging, and fixed in nature. Rituals presumably are actions that occur in a mechanical and fixed manner and, therefore, are not capable of change and modification.

Moreover, rituals are viewed by many as being of secondary importance and epiphenomenal in that they are the product of more important social processes. Other dimensions of society are often presumed to be more "real" and "significant" when it comes to understanding how societies actually work. Finally, all of these assumptions imply that rituals have little effect on or significance for people and developments in society, especially contemporary society. Rituals presumably play a negligible or minor role in human behavior, social interaction, and groups, big or small.

Such a limited approach to ritual usually fails to open up to the full complexity of human behavior. And related to this point many sociologists, while paying attention to social organization, pay little attention to either culture or personality structures (notwithstanding the contributions of those in the sociology of culture, social psychology, and the sociology of emotions) and the ways ritual may be directly related to all three aspects of social behavior.

For these reasons rituals are often thought to have limited explanatory value and are often downplayed in social analysis. In other words, they remain in various ways invisible to and "under the radar" of many students of social life and modern society while other dimensions of social reality, many of which operate at more macro levels, are emphasized in our research and thinking.

Structural ritualization theory (SRT), in contrast, focuses on the role that rituals play in society (Knottnerus 1997; 2005; 2009). It is grounded in the basic assumption that daily life is normally characterized by an array of social and personal rituals. Such rituals help create stability in social life while expressing various symbolic meanings that give significance to our actions. Everyday rituals, whether occurring in small groups or organizations, can lead to consequences unanticipated by group members while both being fed by and feeding into larger societal levels of interaction. As such, this perspective is directed to rituals that occur in various social settings, for example, face-to-face interaction, small groups, organizations, society as a whole, and globally.

Ritual as a Missing Link in Sociological Analysis

More precisely, SRT is based on a very different set of assumptions that enable this approach to provide a missing link in sociological research and analysis. It assumes that:

(1) Rituals are found in both premodern and modern societies. While the presence of rituals in societies prior to the modern era and in so-called preliterate or "primitive" societies is often noted by historians, anthropologists, and other social scientists, SRT also stresses that rituals operate in the modern world. Although the form such rituals may take and the role they play in contemporary society may differ from other cultures in the past, I assume that rituals are a fundamental—that is, universal—component of human behavior and are an integral part of societal processes. The origins of rituals in social life very likely extend into the early development of *Homo sapiens* several hundred thousand years ago (if not earlier than that). They have continued to play a central role throughout history and, I suggest, will continue to do so notwithstanding the profound social changes that have taken place in the world especially in recent centuries.

(2) Rituals occur in both secular and sacred contexts. In contrast to the long-standing presumption among various sociologists and other social scientists that rituals primarily or exclusively involve sacred or religious social practices, SRT assumes that rituals occur in a wide array of settings ranging from the secular to the religious. Rituals may involve standardized behaviors that occur in daily social interaction or more dramatic collective events.

Such behaviors could include, for example, ritualized styles of interaction among people in a school or government office that are, for instance, of an equalitarian or hierarchical nature; religious or civic festivals; dining practices within a subculture; dancing and other stylized physical activities in settings ranging from religious gatherings or political celebrations to concerts or raves; ritualized behaviors expressing status distinctions among slaves, office workers, or youth groups in schools; ritualized deviant behaviors that occur in bureaucratic organizations such as corporations or health care facilities, sporting events, and so on. To restrict rituals to only sacred/religious contexts is to ignore the powerful and vibrant role rituals play in much of our lives and the ways they can meaningfully enrich our behavior, relations with others, and social events.

(3) Rituals are dynamic in nature and subject to change. Oftentimes rituals are depicted as behaviors engaged in by people who are fixed and

static in nature. The image often conjured up in the minds of sociologists and people in general is that rituals are simply repetitious actions mindlessly engaged in by individuals who possess little or no capacity for modification, large-scale change, and/or deliberate (or unintended) invention. SRT assumes that rituals, while certainly involving standardized behaviors, are not inherently rigid or devoid of meaning. Many factors ranging from situational conditions to personality differences can influence the extent to which ritualized actions may or may not be altered. In other words, rituals are inherently dynamic and possess the potential for change.

Indeed, I would argue that a careful examination of history or developments in contemporary society quickly reveals how rituals can change and do so for different reasons and to differing degrees. As examples of such changes one could point to transformations over the last century in major sporting events such as the Olympics; the alteration of national holidays, celebrations, and festivities by various authoritarian governments throughout the twentieth century in order to generate support for their political regimes; modifications in the rituals engaged in by staff in a formal organization due to changes in policy; or the invention and alteration of rituals such as special ceremonies and commemorations by crew members during long-term expeditions (several of these will be discussed more fully in subsequent chapters).

(4) They can be of profound importance in social life. Rather than assuming that rituals are epiphenomenal and reflect more "substantive" and "important" social forces such as organizational structures involving economic, political, occupational, or demographic factors, SRT emphasizes the critical role rituals play in society and how they are a crucial aspect of people's cultural life, interpersonal relations, and the group dynamics operating in the many social environments most individuals move through in their daily lives. In saying this, SRT in no way denies the validity or significance of other social forces. Quite the contrary—this approach argues that while ritual is an essential dimension of social behavior it may both be influenced by these other dimensions of social reality and also profoundly impact them.

(5) Rituals have great explanatory value. This perspective assumes that by attending to the role rituals play in social events we increase our capacity to explain and to better understand human behavior and social dynamics. Indeed, by bringing ritual dynamics "more to the surface" in our observations and analyses of social behavior we are provided with another lens through which to view ourselves. Such a development can

significantly contribute to not only the scientific analysis of society but also our ability to better understand our own personal experiences.

In these different ways ritual provides a missing link in sociological thought. The concept of ritual addresses key issues and arguments such as these, which are often ignored or underemphasized in sociology and other social sciences. What is more, at least six additional meanings or assumptions related to the concept of "missing link" are of particular importance. All of these are relevant to the first five points addressed by SRT while at the same time going beyond them to underscore other fundamental issues. They are as follows:

Rituals occur at and impact different levels of society ranging from face-to-face interaction and relationships, larger groups and organizations, society as a whole, and the world. Ritual operates at micro and more macro levels.

The concept of ritual can be linked to perspectives and traditions of research emphasizing other social dynamics and issues; therefore, approaches focusing on ritual should be capable of forming linkages or conceptual bridges with other perspectives (what is often referred to as theory integration).

Ritual is a social phenomenon that can be investigated with very different methods and types of evidence.

Rituals can have consequences for social organization, culture, and personality structure.

The idea of ritual can provide a common vocabulary and framework to study developments occurring in different groups. In other words, it can have great analytical value helping us to explain the workings of society.

Ritual is a concept that has potential relevance for the multifaceted nature of social life. In other words, ritual can be utilized in a wide range of studies given the complexity of human behavior.

These six additional points are also germane to SRT and its attempt to make ritual an important link in sociological analysis. How SRT addresses these concerns is highlighted in points 6 through 11.

(6) SRT addresses these specific concerns because it examines some of the ways in which social dynamics operating at one level of analysis influence group processes occurring at other levels. This approach assumes that ritual dynamics occur at different—micro to macro—levels of analysis and that ritual processes operating in social environments can influence other social units. For instance, face-to-face interaction in small groups can be influenced by the organization the group is located

in or even the larger community. A six-level model of the social order to be outlined in chapter 2 facilitates the study of ritual dynamics at various levels of analysis. The theory argues that ritual dynamics in a group can spread to and influence ritual enactments in other groups and social milieus both across levels and within each level.

(7) SRT creates different kinds of linkages with other perspectives and areas of study focused on various topics. For example, SRT complements and/or has developed links or bridges with the web/part-whole approach (developed by Bernard Phillips and Tom Scheff), expectation states theory, macro social change/Dark Age research, in addition to studies of organizational deviance, emotions, ethnic communities and identity, disasters, social power, movements, and social inequality.

(8) It uses multimethod research strategies. Research employs multiple methodologies involving both quantitative and qualitative types of evidence, including historical-comparative analyses, experimental research, contemporary and historical case studies, interviews, field research/ethnographies, content analysis of primary sources, reviews and analyses of secondary sources, surveys, focus groups, and interpretations of literary sources and accounts involving novels, autobiographies, biographies, memoirs, travelogues, and diaries. Such an approach implies a respect for the strengths of different methods and recognition of how diverse research strategies can complement one another and enrich our investigations of social reality. In other words, this perspective assumes that when similar findings are obtained employing dissimilar methods focused on social phenomena occurring in different settings and historical periods we increase our confidence in the value of the explanatory formulations.

(9) In addition, SRT emphasizes the relevance of rituals for *social organization* (or *social structure*). It stresses how ritualization occurs and leads to the patterning of social relationships and, therefore, the formation, reproduction, and sometimes alteration of social structures. At the same time, it is directly relevant for understanding *culture* due to the central focus on rituals and the symbolic meanings expressed through them and the role rituals play in providing meaning, direction, and focus to social life and group interaction (concepts and points usually recognized as important by sociology of culture researchers and social anthropologists). Furthermore, this approach has relevance for *personality structures* because it argues that the rituals developed by individuals (oftentimes derived from or influenced by groups) can have profound consequences for people's cognitions, feelings, and overall character. While much

remains to be done in this area, evidence from studies of concentration camp internment shows, for instance, how important personal and group rituals can be for enabling people to cope with highly disruptive experiences and maintaining a stable personality structure. Such an issue I would propose is largely a void in sociological research.

(10) SRT uses a common analytical framework employing the idea of ritual to explain various social experiences. It provides abstract formulations that address basic social processes involving ritual dynamics in many empirical settings, for example, schools, corporations, ethnic communities in urban areas or small towns, slave societies, health care facilities, sporting events, youth groups, political or religious groups, and social movements.

(11) Finally, SRT provides a missing link in sociology because, given the enormous complexity of human behavior and our failure to address that complexity with our specialized approach to research in sociology and other social/behavioral sciences, it is a concept that can be used to examine many aspects of social reality. As just noted, SRT provides abstract formulations and an analytical framework that can be applied to many different phenomena in a wide range of studies. Indeed, while a number of topics have already been investigated, some of which will be discussed in this book, the concept of ritual could be used in a much wider range of studies. Furthermore, given the complexity of human behavior and the limitations of specialized research, ritual may be a useful additional component in investigations of many other research issues.

The openness to different research strategies, a willingness to create conceptual linkages with other perspectives, a concern with different levels of the social order and dimensions of human behavior, and the development of a common conceptual framework make it possible for SRT to study and analyze an extremely wide scope of social groups and events.

Theory Development and Cooperation

To accomplish these objectives this approach builds on the contributions of scholars concerned with ritual such as Durkheim, Goffman, Warner, and Collins and other analysts such as Berger and Luckmann (1966; see Knottnerus 1997; 2009 for a fuller discussion of influences). However, SRT's ideas are presented in a more formal manner than are the arguments of most of those who have contributed to our understanding of ritual or related issues. SRT more formally defines ritualization and presents a set of factors to explain various dynamics involving social action and social structure, which will be discussed in chapter 2.

The theory focuses on the role that "ritualized symbolic practices" play in social life and the ways they contribute to the creation, reproduction, and/or transformation of social structure. Being formally stated with precisely developed concepts, SRT exhibits a necessary quality of predictive theory. Indeed, the development of precise concepts facilitates the use of the framework for systematic investigations of empirical cases and a clearer understanding of the processes occurring in these social settings. In essence, the theory strives for cumulativity, that is, a growing corpus of knowledge about how ritual influences social behavior and the workings of human societies.

I would also emphasize that my approach to the idea of linkage in regard to sociological analysis involving ritual is influenced by the basic assumption that knowledge production is advanced by a spirit of cooperation in theory and research—not just competition (Knottnerus 2005; see also Sell 2007; Knottnerus 2007). Theory, in its most fundamental sense, furnishes us with a conceptual framework for understanding and explaining social reality. It provides a logical set of principles that help us to comprehend social life and that guide our research practices. It is the abstract "lens" through which we seek to make sense of the myriad social behaviors and events in human experience.

Of course, the quest for theoretical principles and research evaluating these ideas, as I have noted elsewhere, is conducted by people who are themselves social actors influenced by the various objectives and pressures imposed on them by their professional discipline, the academic organizations within which they teach and conduct research, and other conditions shaping human behavior including the desire for a positive social identity (Knottnerus 2005:6-7):

> Perhaps one of the most prominent features of these conditions surrounding us is that the development of theoretical ideas and related research takes place in a competitive social environment. We work in a competitive professional milieu which values knowledge produced through a spirited and vigorous comparison between and selection of those approaches that appear to garner the greatest empirical support and impress us as being the most persuasive. This is in many ways a desirable situation. However, as is the case for many things in life, such a situation can also go too far and have less than positive consequences. Competition when carried to

> extremes can have undesirable consequences in the world of theory and research. Competition can lead to insensitivity to the merits of alternative perspectives. It can lead to a disinclination to appreciate how other approaches can contribute to our understanding of the world. And, perhaps even worse, it can lead to a type of professional one-upsmanship and intellectual imperialism motivated to varying degrees by the desire to confirm our own professional identities while negating competing viewpoints and other potentially useful research.
>
> For this reason, I suggest that such tendencies should be countered by recognition of the value of cooperation in our theoretical and research efforts. Cooperation, rather than competition, can also be a productive conduit through which knowledge production may occur. By seeking to interrelate theories and research findings we are provided with a means for broadening our understanding of and developing new insights about social life.

In supporting such an approach I recognize that various types of theories exist ranging from, for example, general or broadly developed arguments to more formalized presentations. My own preferences point toward the development of abstract concepts and formulations that are precise and systematic in their presentation. By doing so, we create theories that are more clearly understood and capable of being studied, in other words, theoretical concepts that research can provide tests, corroborations, or exemplifications of. Moreover, by striving for clarity we obtain other benefits, including the development of more clearly defined questions about the issue we wish to understand. And to the extent that our ideas are more clearly developed we should be better able to achieve the goal of greater cooperation.

Cooperation can occur in different ways. Furthermore, others have in different ways advocated for or accomplished these goals. For instance, one could point to the arguments and contributions of Bernard Phillips (2001), Jonathan Turner (2003), Randall Collins (1975), and certain researchers conducting investigations in the network exchange tradition (Willer 1999; Lawler and Ford 1993; Whitmeyer and Cook 2002). Cooperation could involve different theoretical approaches linking or integrating their concepts and principles. Or, it might involve conceptual bridges between perspectives, that is, establishing areas where perspectives may interrelate and complement one another.

The logic of cooperation in theory also has relevance for the research strategies we use. Rather than emphasizing a competitive relation between different research strategies there is much to be gained from embracing different methods and kinds of evidence, what some refer to as multimethod research. Such an inclusive orientation provides a richer array of evidence that can be used to corroborate and illustrate our theoretical ideas. The bottom line, however, is that we stand to gain from an approach focused on cooperation rather than one solely driven by a competitive stance.

SRT's commitment to cooperation and integration is realized in various ways. As already noted, SRT has created conceptual linkages with a number of other perspectives. It also uses multimethod research strategies. Moreover, it gives explicit attention to the ways social dynamics involving ritual at one level of the social order influences group processes at other levels. Finally, SRT uses the concept of ritual to examine social behavior, relations, and structure in quite different groups.

In saying this, let me underscore how SRT focuses on rituals in daily social life. What is of ultimate concern to this perspective are the taken for granted, daily ritualized practices of social actors, which both influence other groups and social processes operating in society and may be influenced by those other social environments and conditions. This relates to the aforementioned point (number 6) concerning, for instance, how ritual processes operating at more macro levels can ultimately influence the interpersonal behaviors of people in their daily lives and vice versa.

I should also note that although the theory addresses both personal and social rituals and recognizes that in certain situations some rituals may be carried out individually, SRT has emphasized the *social* nature of rituals. Whether rituals are conducted by one or more actors, this perspective stresses how they are not infrequently derived from sources outside of individuals, may develop through interaction with others, oftentimes have social referents, and express meanings or themes that are collectively understood by and are significant to others. It is still important to realize though that in terms of their enactment rituals may be engaged in individually or collectively. Consequently we can talk about "individualized ritualized symbolic practices" (noting the above points about the social antecedents and qualities of such personalized rituals) and socially shared rituals, that is, "collective ritualized symbolic practices." This is a distinction that parallels Bernard Phillips's conceptualization of rituals and social rituals.

In focusing on rituals in daily social life, SRT differs from many other theories and perspectives. For instance, unlike social exchange theory (Cook 1987), game theory (Bienenstock and Bonacich 1993), and elementary relations theory (Willer and Anderson 1981; Willer 1999), SRT is not driven by rational choice. While sharing some assumptions with social psychological perspectives such as symbolic interactionism (Blumer 1969), expectation states theory (Berger, Fisek, Norman, and Zelditch 1977), and social cognition theory (Howard 1995), it differs from these traditions due to its explicit focus on ritual enactments, its willingness to employ different methods and types of empirical evidence, and a concern for social dynamics occurring at different, more macro levels of the social order. And while concerned in part with the social dynamics contributing to social inequality and power, it also differs from approaches such as Marxist theory (Marx 1967; 1970) and world-systems/social change perspectives (Wallerstein 1974) because of its emphasis on the symbolic and meaningful components of social life and culture. SRT addresses a different kind of social phenomenon than theories such as these.

In essence, this approach rests upon the assumption that ritualization is a key dimension of social behavior as are other aspects of social action such as rationality emphasized by social exchange theory and rational choice theory, symbolic interpretation by symbolic interaction, or emotions within the sociology of emotions. Put somewhat differently, *ritual is like an engine that drives much social life,* sometimes quite intensely.

Goals of SRT and Related Theory and Research

All of the work in SRT centers on three interrelated goals: (a) the development of abstract theoretical ideas explaining ritual dynamics; (b) empirical research that substantiates and illustrates these abstract concepts; and (c) applying this knowledge to alleviate social problems, social inequality, and dehumanizing social conditions.

These goals complement and are inspired by others who have emphasized the need for the development of abstract concepts to explain social dynamics, the importance of empirical methods to study and corroborate those ideas, and utilizing this knowledge to address social and personal dilemmas confronting humankind in contemporary society. Of particular importance in this regard are the arguments of C. Wright Mills (1959), which were put forth most persuasively in his famous work *The Sociological Imagination,* and the writings of Bernard Phillips (and others such as Tom Scheff), who has emphasized how such an approach is essential to our understanding of the complex nature of social behavior and solving the many social problems confronting us in an increasingly dangerous and complex world. The latter's ideas—referred to as the web/part-whole approach—have been presented in a number of books published in the series Advancing the Sociological Imagination with Paradigm Publishers (Phillips 2001; 2007; 2009; Phillips and Christner forthcoming; Phillips and Johnston 2007; Phillips, Kincaid, and Scheff 2002; Knottnerus and Phillips 2009).

I should also stress that while not necessarily identifying Mills as an intellectual influence other research traditions and scholars in sociology have to varying degrees highlighted the importance of theory, research, and application and have, therefore, contributed to the goals that shape SRT. A prime example would be certain theoretical/research programs in social psychology and group processes such as expectation states theory and related work emphasizing the importance of a generalizing orientation in sociology (Berger, Fisek, Norman, and Zelditch 1977; Berger and Zelditch 1993; 1998; Szmatka, Skvoretz, and Berger 1997). Of course, others have also stressed in different ways the importance of abstract theory and empirical research in the sociological tradition ranging from contemporary analysts such as Jonathan Turner (2003) to some of the classical scholars who established the foundations for the discipline such as Emile Durkheim and Karl Marx.

In this book attention will be directed mostly to theory and research dealing with ritual dynamics. The last two chapters (especially chapter 7), however, will directly address the need for application, interventions, and policy development; briefly describe recent research addressing problems found in contemporary society; and discuss some personal and social implications derived from the theoretical ideas and research findings of SRT.

Reflecting SRT's commitment to address the enormous complexity of human behavior, this analytical perspective has been used to examine many different phenomena in an increasingly wide range of studies. Others and I have carried out, or are in the process of conducting, a number of investigations of ritual dynamics in different social milieus. Studies focus on ritualization occurring in both contemporary and historical settings. This research also examines ritual enactments in American and international settings. In doing so, the theory shows that it can bridge across methods that have been applied in the field, in historical research, and in the laboratory. SRT is unique in the very broad scope over which it has been applied.

Many of these studies involve multiple researchers. In other words, a large amount of this work is collaborative in nature involving several persons or what might be referred to as teams of investigators. These individuals and the different research topics they have addressed or are addressing will be formally identified, that is, referenced at appropriate points throughout the book. This approach to conducting research rests on the assumption that social science can be productively advanced through cooperative endeavors. While individual effort and insight are crucial parts of science, at the same time, research is advanced through collective projects and the stimulation and insight that emerge from dialogue and intersubjective ties and communication.

Over the last several years SRT's analytical focus has also grown. Several lines of development are currently under way, each of which builds on and involves an extension of the original formulation dealing with structural reproduction, which will be discussed in the next two chapters. And a number of studies have been or are currently being carried out providing tests, exemplifications, and applications of the original formulation and these extensions. All of this work, to repeat, is committed to the development of abstract concepts at varying levels of abstraction, grounding these concepts in empirical evidence, theory integration, and using this knowledge to mitigate social problems.

In this book I focus on three specific directions of research: structural reproduction, strategic ritualization and power, and disruption and deritualization. Later, in chapters 6 and 7, I will identify other developing lines of research.

Before doing so, however, I need to reiterate that SRT focuses on everyday life and the central role rituals play in our personal and social behavior. Indeed, these ritualized practices, which we often take for granted and do not question, define in many ways the very nature and quality of our lives. The study of ritual enactments in our everyday activities is, therefore, of crucial importance and can provide various insights into our behavior and social relations. These points are of direct relevance for analysts, students, and all of us whose lives are both consciously and unconsciously shaped by the complex, powerful, and not always constructive social forces operating in contemporary society.

Chapter 2

Analysis of Ritual Dynamics

In its original formulation (Knottnerus 1997) SRT emphasizes the importance of embedded groups or groups that are nested within a more encompassing collectivity. Examples of such groups would include an informal youth group in a school, a slave society within a plantation, or a problem-solving group in a bureaucratic organization. The formulation focuses on the ritualized actions performed in a wider social environment that acquire significance for the actor and then become part of the individual's script for his or her immediate world. In this way, although practices are not just "copied," they surface and are expressed in ways that may confirm the pattern of activity in the wider social milieu. Ritualized practices develop in ways that confirm the patterns of behavior in the wider social environment, that is, they are reproduced.

In this regard, SRT highlights an irony of social behavior that the reproduction of social structure occurs even when it may not be in the interest of people and the groups they are members of. Such reproduction includes, for instance, egalitarian problem-solving groups assuming the hierarchical and more rigid features of corporate organizations they are located in, children from abusive families becoming adults who abuse their own family members, or the mirroring of status distinctions among owners of slave plantations within the slave society itself. Crucial to all of these developments is the central role that ritual enactments play in daily social life.

Structural Reproduction

More precisely, the theory focuses on the role that ritual dynamics play in the creation and reproduction of social actions and social organization in groups embedded within a larger social milieu. According to the theory, certain elements play a major role in the ritualization processes that produce behaviors, interaction patterns, and structures similar to or different from those taking place in the larger social environment.

An embedded group refers to a group located or nested within a more encompassing collectivity such as a social institution, organization, or simply a larger group—for example, a faculty committee within a university department, an informal youth group within a school, or a group of friends within a club or a musical group. Berger and Luckmann's (1966) social phenomenological approach provides a general perspective for better understanding how embedded groups can be swayed by the wider environment. From their point of view the social organization and ritualized behaviors of individuals in the wider social milieu would constitute an objective reality for actors within the embedded group. They are the predominant, taken-for-granted patterns of behavior and meanings internalized by the members of such groups in their cognitive scripts or typifications of the social world. They are then externalized or re-created in their immediate group and relations, that is, they become the ritualized behaviors that shape face-to-face interaction. The authors (and I) assume that people use the symbolic and social resources they are exposed to in their construction of new behavioral patterns. While they are not simply copies of the larger social environment these newly constructed rituals become solidified and enduring social arrangements, that is, objectified or reified. Once developed, the rituals and social organization of the embedded group directly impact the thoughts and acts of its members. These re-objectified and reified structural forms then reproduce the actions and social patterns of the larger social milieu. Whether the embedded group is similar to or different from the wider environment is influenced by the nature and arrangement of ritualized activities in the larger setting.

In presenting a more clear-cut explanation of these processes than is provided by Berger and Luckmann's phenomenological perspective, the concept of ritual clearly plays a pivotal role. While building on the contributions of various important scholars who have examined ritual (as previously noted in chapter 1) I also attempt to present a more precise discussion of ritual dynamics than most of these analysts do. I would also emphasize that while the idea of ritual is sometimes used only to discuss religious, ceremonial, and sacred behaviors, I use the concept in a broader sense to denote activities such as ritualized behaviors and social interaction that occur in all kinds of contexts, including secular settings. This orientation is quite consistent with Kertzer's (1988) treatment of ritual as socially standardized and repetitive symbolic activity.

Moreover, the arguments of Giddens (1984) and Sewell (1992) and their discussion of agency and structure inform parts of the current formulation. SRT assumes that both agency and structure are integrally involved in the workings of society and the dynamics at the heart of social development, transformation, and maintenance. Giddens has also emphasized how routine plays a central role in social interaction and institutionalization in society. While not exactly the same, his conceptualization of "routine" and the stress placed upon it exhibits a general similarity to SRT's approach to ritualization and its importance in daily life.

Finally, some of the processes addressed in the first formulation of SRT parallel Simmel's (1906; 1950) observation that secret societies often possess hierarchies and rituals and these are derived from the social world surrounding them. Indeed, the resulting inner world of a secret society may be very similar to the external society even if it stands in opposition to it. Put somewhat differently, to produce something new requires using something old. So while agents are autonomous and active, they still have to employ symbols, resources, behavioral models, and so forth from their prior experiences and surrounding environment in the process of social construction. In so doing, they may reproduce a preexisting or previously objectified social reality.

Building upon these different scholarly contributions, the concept of "ritualization" is used to explain how social structure is reproduced and changed. The theory, therefore, directs attention to how social dynamics operating at one level of society influence what occurs in groups at more micro levels. Because SRT focuses on the particular analytical concerns that it does distinguishes it from most of the aforementioned theories and research. Moreover, the theory in its original form is presented in a much more formal manner than those other discussions of ritual, some of which are indifferent if not opposed to theoretical formalization. Many (includ ing admirers such as myself) view certain of these approaches as lacking in specificity, vague, and hard to investigate or test (e.g., the writings of Giddens or Berger and Luckmann). More formal definitions of rituals are developed in SRT, which creates a more precise theory. Such an approach requires that arguments be presented in a logical manner with clarification of theoretical principles, assumptions, and orienting frameworks. The formal development of concepts and arguments makes it easier to conduct research that addresses the theory. It also means that further directions for theory development and research may be more clearly delineated.

Stated somewhat differently, a number of concepts have been developed that allow for a more precise examination of empirical evidence and tests, exemplifications, and applications of the theory. Indeed, SRT presents precise definitions and concepts and clearly defined variables that can be operationalized and employed in research, whether qualitative or quantitative in nature.

The present discussion will, however, be less formal (for the more complete presentation of the formal theory, see Knottnerus 1997). Basic definitions and concepts will be presented as they appear in the theory. These principles will then be discussed more informally, and various examples and studies that provide vivid examples of how these principles play out in real life, past or present, will be presented throughout the book.

In essence, the theoretical discussion to follow argues that when ritualized practices in a social environment are strongly accentuated and presented to the members of an embedded group they generate "schema-driven action repertoires." The greater the impact of these ritualized activities on individuals the more similar will be their immediate social world and the wider environment within which they are located. When, on the other hand, people are exposed to ritualized behaviors that differ from one another, the organization of their embedded group and the wider milieu will be less similar.

Theoretical Principles

The theory examines the role symbolic rituals play in the formation, reproduction, and alteration of social structures and outlines a number of conditions and processes that are sufficient for structural ritualization to occur. It does not, I would emphasize, state the theory in terms of necessary conditions. Rather, it focuses on a set of sufficient conditions for the occurrence of certain kinds of ritual dynamics.

Before defining key concepts in the theory, several terms should be clarified. "Socially standardized" refers to a social practice that is regularly engaged in. It implies a (more or less permanent) continuance of a social usage or what is often thought of as a custom or a customary way of doing. That said, the first concept, *action repertoire,* may be defined as a set, the elements of which are socially standardized practices.

Another important term is *schema,* which may be thought of as a cognitive structure. In a similar vein, schemas are also depicted as cognitive frameworks or structures (or organizations) of conceptual representations dealing with items such as situations, events, objects, sequences of actions and/or events, and so on (for discussions of schemas in social psychology, see Markus and Zajonc 1985; Fiske and Taylor 1991; and for discussions of how similarities exist between many ideas in phenomenology and cognitive social psychology—e.g., schemas—see Forgas 1981). With these presentations of the terms *action repertoire* and *schema* the principal concept of the theory can be developed.

Ritualized symbolic practice (RSP) is an action repertoire that is schema driven. Ritualized symbolic practices refer to the ubiquitous form of social behavior in which people engage in repetitious and regularized actions that are grounded in individuals' cognitive maps or, to phrase it somewhat differently, symbolic frameworks. Such ritualized activities occur throughout social life and include among other things regularized styles of interaction within various organizational settings such as schools, businesses, hospitals, or governments; prevalent types of relationships within informal groups, subcultures, or secret societies; commonplace social practices within political, commercial, religious, or civic settings; and stylized bodily movements and vocalizations such as dancing, singing, chanting, or marching.

This perspective emphasizes that rituals inescapably involve two crucial dimensions: actions and cognition. Social behavior in the form of ritual enactments involves cognition, that is, conceptual representations of reality. Or to put it a little differently, RSPs rest on a meaningful or symbolic component. Ritualized practices, therefore, express or communicate to varying degrees symbolic themes or meanings.

In saying this, the theory assumes that not infrequently people engage in ritualized behaviors that involve little conscious attention and that they do not necessarily reflect or think about. Moreover, such ritualized actions may be weakly rational or even viewed as arational (others who have made similar arguments about human behavior in general include Camic 1986; Ralph H. Turner 1991).

All the arguments of the theory in its first formulation described here apply to certain scope conditions or contexts designated as "domains of interaction." A domain of interaction is a bounded social arena that contains two or more actors. These individuals are at least part of the time engaged in face-to-face interaction. Furthermore, a domain of interaction involves a delimited region or sphere of social behavior. This social arena has the power to produce effects, meaning it affects the probability of occurrences. These occurrences involve people's cognitions and actions.

This concept points to the fact that single or multiple domains of interaction may influence the development of ritualized practices. In other words, we can distinguish between social environments such as organizations or institutions in which embedded groups may be exposed to RSPs in different numbers of interaction domains for varying lengths of time. For instance, in one organization an embedded group might come in contact with certain types of ritualized practices in eight domains of interaction for a total of nine hours a day while in another setting a similar group might be exposed to such practices in only one domain of interaction for one hour. Differences such as these could be of great importance in terms of these environments' effects on groups embedded within them, that is, the extent to which RSPs similar to those found in the larger milieu emerge within these groups.

To denote a setting such as an organization that comprises multiple domains, the concept of "domain set" has been developed. Here, the idea of *domain* is partly used in a mathematical sense, for example, social settings that may contain different numbers of social arenas in which people encounter certain kinds of ritualized practices. But domains can be further identified in terms of their characteristics, functions, and so on (e.g., the domains of the dormitory, classroom, and play area in a boarding school or the domains of the dining room, physical therapy room, outdoor walking area, and resident's own room in a nursing home).

Having defined the aforementioned concepts and conditions we can now identify the factors that influence the dominance of RSPs in a larger social milieu. Four factors are formally stressed in the theory.

Salience refers to the degree to which an RSP is perceived to be central to an act, action sequence, or bundle of interrelated acts. As already emphasized, ritualized actions contain symbolic schemas or models. Symbolic meanings are grounded in and expressed or communicated through the ritualized practice. Salience, therefore, refers to the extent to which an RSP within a domain of interaction is conspicuous, prominent, or noticeable. What is suggested is that ritualized practices can vary in the degree to which they are salient.

In the original formulation of the theory the example of handshaking versus interaction between staff and students in nineteenth-century

French male lycées illustrates how salience can vary. In American culture handshaking usually is a ritualized behavior that has fairly low salience. It is carried out in a routine fashion and is not central to the dynamics of interaction among different people. It doesn't permeate and distinguish the behaviors and relationships among interactants. On the other hand, certain ritualized behaviors engaged in by staff and students in the French schools were extremely salient (as will be described more fully in the next chapter). In these settings teachers and supervisors always exercised strict discipline, dealt with students in an authoritarian manner, were extremely demanding, and were quite distant in their demeanor and relations to youth. These formal, hierarchical relations were quite prominent in all their social behaviors. Such practices were highly visible and conspicuous to all parties, infusing and shaping most aspects of the interaction between the two groups from the dynamics of the classroom to the dormitory, that is, different social domains of interaction. These ritualized actions and the symbolic meanings expressed through them were highly salient.

Repetitiveness is defined as the relative frequency with which an RSP is performed. This component of the theory addresses the basic fact that ritualized behaviors can vary in the degree to which they are repeated. The spectrum of occurrence can range from a practice rarely occurring in a domain of interaction to its being continually engaged in. For instance, in certain institutions distinctive ritualized practices may happen hundreds of times a day, while in other settings they may be an infrequent occurrence. A simple example would be a school where staff repeatedly engages in behaviors with students that could be described as informal, highly personal, and affectionate. Yet in another educational institution a more impersonal and authoritarian tone might mean that displays of affection hardly ever occur. The degree to which RSPs are performed by actors can have profound consequences for the social dynamics and pattern of relations that form in groups.

Homologousness refers to the degree of perceived similarity among different RSPs. This component focuses on the similarity of the ritualized practices occurring in a domain of interaction or cluster of domains within a social setting. Ritualized practices may exhibit to differing degrees a perceived correspondence in their meaning and form. The greater their correspondence the more likely they will have the same effect or outcome. In other words, these behaviors reinforce one another, increasing their dominance in the environment and their potential impact on individuals within an embedded group.

For example, higher degrees of homologousness are more likely to be found in settings such as traditional elementary classrooms where the teacher possesses high degrees of authority. Social practices such as children raising their hands to ask questions, students asking permission to get a drink of water or to use the restrooms, children being seated on command or lining up, and students having their playtime regimented by teachers are different ritualized practices, but are similar or homologous to one another in the teacher's and children's constant status positions and their hierarchical structure. In contrast, less traditional classrooms might be characterized by the ritualized activity of children raising their hands to provide answers or to ask questions, but might have no rules about use of drinking water or restrooms and students organizing themselves for movement from one location to another (as opposed to lining up or sitting down on command) or where and what to play during recreational periods. The social domain or arena of the traditional classroom would, therefore, have a higher level of homologousness than would the less traditional classroom setting.

A further conceptual distinction can also be made between two types of homologous ritualized symbolic practices (HRSP). "Concurrent HRSP" refers to HRSPs that occur within an immediate domain of interaction. In this case, ritualized activities are simultaneously present in a social environment. "Successive HRSP" refers to HRSPs that occur within a domain of interaction at different temporal or historical periods. Here, HRSPs can operate within a social sphere in different generations or time periods and still have a powerful impact on people. Generally, this discussion suggests that HRSPs enhance the reproduction of social structure in embedded groups. Nonhomologous practices, in contrast, do not have the same effect. Instead, they are more likely to lead to innovation, change, and new social patterns. I will elaborate on this idea later in this chapter.

RSP *resources* are defined as the materials needed to engage in RSPs that are available to actors. This factor stresses the importance of resources for ritualization to occur among actors in social arenas. The greater the availability of relevant resources the more likely or easier it is for ritualized practices to be engaged in by people. This analytical factor emphasizes the crucial point that while ritual enactments are predicated on cognitive or symbolic representations of social reality, ritualized practices are accomplishments by corporeal agents who need resources to act in the world. Scholars such as Giddens (1984) and Sewell (1992) highlight this basic point in their discussions of how rules/schemas and resources are at the core of social action and the "duality of agency and structure."

Resources can include many kinds of attributes and commodities. They can include social skills, cognitive and intellectual abilities, knowledge, physical strength, organizational supplies, musical instruments, clothing items, and so on. Moreover, actors vary in their capacities to engage in RSPs and to be influenced by the ritualized activities they come in contact with. Some practices may require resources such as behavioral skills or cognitive abilities that people do not possess. Only those practices that can be managed by individuals will be used by them. For instance, the learning of a particular type of dance step or the playing of a musical instrument may require physical and mental abilities that an adolescent but not a younger child possesses.

A distinction can also be made between two types of resources. "Human RSP resources" refer to the abilities and characteristics of actors perceived by group members to be of value (or to have utility) for themselves or the group. Resources of this type could involve attributes or abilities of individuals such as technical expertise, verbal skills, cognitive abilities, managerial skills, status cues, or occupying a position of authority in an organization. "Nonhuman RSP resources" are defined as all that is not human that is perceived by group members to be of value (or to have utility) for themselves or the group. Examples here might include computers, money, school supplies, physical props and equipment, banners, or food. These distinctions and examples indicate how varied and important resources are for carrying out different kinds of ritualized practices in different situations. Ample evidence of these differences and the significance of resources will be evident in the subsequent chapters, which examine different kinds of ritual dynamics.

In addition to these four factors one other definition must be presented. These four components influence the rank or strength of RSPs in a social milieu. *Rank* is defined as the relative standing of an RSP in terms of its dominance. The theory emphasizes that ritualized practices can differ in their dominance or standing in a domain of interaction. They can vary in their strength or importance with a particular social arena. Moreover, RSPs can be assigned a value to indicate their degree of influence or dominance.

Various assumptions and additional definitions are presented in the theory to explain the processes involved in structural ritualization. First of all, the rank of RSPs in a social environment is assumed to be the sum of salience, repetitiveness, homologousness, and resources. The greater these factors, the higher the rank of those ritualized activities in the domains of interaction that people are exposed to and/or engage in. In the original version of the formulation and in subsequent research these four factors are weighted equally. I would point out that it is possible that the value of some or all of these factors could vary under different circumstances and so they could be modified in more developed versions of this formulation. The key point, however, is that the higher the rank of RSPs in terms of salience, repetitiveness, homologousness, and resources, the greater the strength or dominance of these behaviors, that is, the greater their effect, influence, or importance.

This condition directly impacts the subsequent steps in the ritualization process. Stated in more general terms, the higher the rank of ritualized activities in a more encompassing social milieu or environment, the greater their impact on the thoughts and deeds of individuals who are in contact with these practices and who create informal groups of their own nested within this larger social environment.

More precisely, the theory assumes that RSPs shape actors' schemas in embedded groups. Dominant ritualized practices in a social milieu influence the schemas or symbolic models developed by actors in nested groups. Dominant or important ritualized practices are significant sources of meanings that people utilize as they construct their own cognitive schemas.

Once formed, I assume that schemas shape actors' action repertoires in embedded groups. In other words, these newly developed schemas or cognitive structures form the basis for and influence individuals' actions within their group. Individuals, therefore, are more likely to engage in these new behavioral performances due to the underlying cognitive or conceptual framework. As actors increasingly engage in these practices they become regularized or standardized and eventually patterned modes of behavior within their immediate domain of interaction. In essence, these regularized or repeated action sequences become new RSPs engaged in by individuals within their group.

At this point another definition should be provided. *Structural isomorphism* is defined as the equivalence of social relations and their pattern of organization in a social environment and an embedded group. This definition refers to a situation in which social relationships and structure in an embedded group are similar to the social relations and pattern of organization in a larger social environment.

Here, it should be noted that elsewhere Christopher Prendergast and I (1993:160) have defined social structure as "... a domain of phenomena consisting of social relationships, the opportunities for and constraints on joint action they entail, and their coalescence into complex systems of different types." And in a more recent work, we define social structure as "social relations of various kinds, their internal processes, their behav ioral and institutional issue, and their connectivity and interdependence as emergent systems extended across time and space" (Prendergast and Knottnerus 1994:1). Among the various features referred to in these definitions, probably the most important ones emphasize that structure involves a system of social relations. This is quite consistent with SRT, which conceptualizes structure as "social relations and their pattern of organization." In the present formulation, we are concerned with the structure of social relations involving RSPs, which by definition involve regularly engaged-in social practices (for a useful review of analytical discussions of social structure, see Bernardi, González, and Requena [2007] in the comprehensive two-volume set, *21st Century Sociology: A Reference Handbook*).

Having defined structural isomorphism, we can present the last assumption dealing with structural reproduction in embedded groups. It states that the greater the rank of RSPs in a social environment, the greater the structural isomorphism between ritualized practices in the embedded group and the environment. In other words, the higher the rank or dominance of RSPs in the wider social milieu, the more equivalent the pattern of ritualized behaviors and relations in an embedded group and the larger environment.

To repeat, the greater the rank of RSPs in a larger social milieu (in terms of the previously discussed factors), the more likely ritualized practices in the embedded group will be comparable to those activities in the wider environment. Due to the impact of these dominant action sequences on people, the schema-based action repertoires developed by individuals in an embedded group result in ritualized practices that are similar to those occurring in the larger milieu. This development results in a structural isomorphism between the nested group and the more encompassing social environment. The nested group contains an array of RSPs and relations that are comparable to the social arrangements—that is, ritualized actions, patterns of behavior, and type of social structure found in the surrounding, larger domains of interaction.

In sum, the greater the degree of salience of RSPs, repetitiveness of ritualized activities, presence of homologous ritualized practices, and availability of RSP resources, the higher the rank of RSPs in the larger environment. The greater the strength or dominance of these practices, the more likely new, yet comparable, RPSs will emerge among individuals in a nested group. Due to the similarity of ritualized behaviors and patterns of relations in the wider social setting and the embedded group, isomorphic social structures will result. In making this argument I would stress that while these ritualized enactments and relationships are comparable in form or structure these new arrangements are reworked and constructed by agents. People use the symbolic models and behavioral repertoires that are available to them as they engage in new ritual enactments and the reproductive process.

Ritual Transformation

As important as the reproduction of rituals and social structure is, it is also possible that ritual dynamics involve innovative processes by which new social arrangements may be created. This argument is consistent with the observations of Durkheim (1965; Young 1994:75) who noted that people use rituals not only to reproduce social life but, in certain cases, to create new social forms and relations. The first process is termed "reproductive structural ritualization" and the second one, which is concerned with changes in ritual practices and structures, is referred to as "transformative structural ritualization" (see Hays 1994 for a partially related discussion of the different ways agency and structure are linked).

While the latter has not been formally researched to any great degree and there are various ways rituals can change, the theory in its original formulation does address one way rituals can be altered and what causes such transformations to occur. According to the theory, transformative structural ritualization could occur when individuals in embedded groups are exposed to different RSPs in one or, most likely, multiple environments. The greater the dominance of these ritualized activities, the more likely individuals will be influenced by and will perform ritualized behaviors that incorporate elements of them. Such an integrative or synthetic process would lead to a group structure different from that found in these milieus.

Without going into the formal details of the argument, one definition and an assumption can be presented that capture the essence of this idea. *Structural non-isomorphism* refers to the nonequivalence of social relations and their pattern of organization in a social environment and an embedded group. Here, social relations and their structure in an embedded group are not comparable to the relationships and pattern of organization in the wider, surrounding social environment.

The assumption dealing with structural transformation in embedded groups states that the greater the rank of different RSPs in a social environment(s), the greater the structural non-isomorphism between ritualized practices in the environment(s) and the embedded group. In other words, the higher the rank of dissimilar ritualized activities in the wider milieu(s), the less similar the organized pattern of ritualized practices and relations in the wider environment(s) and an embedded group.

In essence, the greater the dominance or rank of disparate ritualized activities individuals are exposed to or engage in, the more likely actors in embedded groups will employ elements of the practices to construct new schemas and, therefore, RSPs that are different from these highly ranked practices. These schemas, which are at the heart of the newly developed ritualized behaviors, are synthetic constructs. They combine, amalgamate, or unite elements of meaning derived from the various dominant RSPs. And we should remember that this argument builds on many of the previously presented definitions and assumptions because it assumes that the rank of dissimilar ritualized practices results from processes involving salience, repetitiveness, homologousness, and RSP resources.

A simple hypothetical, abstract case illustrates the idea. If because of these processes individuals come in contact with two highly ranked RSPs—*x* and *y*—in two domains of interaction—*p* and *q*—we should expect, *ceteris paribus* (everything else being the same), the development of a new ritualized practice—*z*—in the embedded group. What results from these processes are non-isomorphic structural arrangements. Or to provide a more concrete example, if the members of a premodern, non-Western society are continually exposed to certain ritualized practices and beliefs shared by Western missionaries, but also continue to come into contact with their own native religious rituals, we could likely see the development of a new set of ritualized practices and beliefs that synthesize or combine elements from both sets of RSPs. Such a development has been observed in many corners of the world ranging from the South Pacific to Africa over the last several centuries.

In conclusion, given a lack of isomorphism or similarity between the structure of the embedded group and the larger milieus, the action patterns and social organization of the nested group will not be equivalent to those found in the other social domains. Instead, a new system or arrangement of RSPs develops that is different from those found in the wider social environments.

As previously mentioned, this formulation concerning one way transformative structural ritualization can occur is presented in the original discussion of SRT. I would stress, though, that ritual change can be due to other reasons and can take very different forms. So far limited attention has been given to the dynamics by which rituals and social structure can be transformed. Most attention has been directed to the first part of the theory dealing with structural reproduction. I will return, however, to this issue in the final two chapters and will briefly discuss at least one other way ritual change can occur in certain social arenas, in particular formal organizations such as health care facilities.

Ritual Dynamics and Levels of Social Structure

As work has progressed in recent years new concepts continue to be developed dealing with ritual dynamics. Here, I end this chapter by adding to these concepts by discussing a new idea. This idea emphasizes that social rituals can operate at different levels of the social order (i.e., micro to increasingly macro levels of analysis) and ritualized behaviors operating in a particular setting can influence the rituals that develop in different contexts or levels.

A basic model for depicting levels of social analysis was first presented in Prendergast and Knottnerus (1994) in *Recent Developments in the Theory of Social Structure*. In the first presentation of SRT, I (1997) suggested that this framework could be used for further theory development. A basic aim underlying the first formulation of the theory was to extend the analysis in ways that reflected a commitment to the spirit of theory integration. Such an exercise could generate an orienting framework that would provide the basis for further theory development and research. I suggested that two dimensions of such a framework would allow for more integrative efforts: one focusing on how ritualization occurs across levels and another on how ritualization occurs in conjunction with other structural processes. These ideas, however, were simply broached in this article and not elaborated on. Here, the first concept is formally presented and discussed for the first time.

The original model for depicting levels of social order (Prendergast and Knottnerus 1994) conceives of social structure as composed of six levels. The levels of social structure are relationships, networks, intraorganizational relations, interorganizational relations, societal stratification, and the world system. The scheme is presented as a heuristic device that rests on a basic assumption in structural analysis that structure unfolds into structure without ever reaching something unaffected by structure. In using what Bernardi, González, and Requena (2007:166) describe as a "Chinese box-type metaphor," we "understand social structure as systems of social relations that manifest themselves with different levels of complexity and that maintain among themselves nested ties." As one proceeds from the level of relationships to the world system, the number of nested ties goes from few to many, complexity increases, and the scope of effects increases from a narrow to a wider range. In the words of Blau (1981:12): "social structures are nesting series with successive levels of more and more encompassing structures." According to this approach, "the most complex systems of social relations include the simplest ones, although each level has its own properties and characteristics. Besides, the logic of each level of social structure is not determined by the higher or lower levels of the structure" (Bernardi, Gonzalez, and Requena 2007:166).

The development of such a division of levels involves balancing intuitiveness and simplicity with descriptions of interaction and designations of causal power. Different models are certainly possible and various level frameworks have been presented (Nadel 1957; Warriner 1981). The present model charts the division of labor between traditions of research, while also providing an analytical framework with a theoretical logic or rationale of its own.

In describing this analytic framework the authors emphasize that the levels scheme breaks a complex whole into its respective parts. The parts are systems of social relations connected by social relations. In two ways the social relationship is the fundamental unit of analysis:

> First, it is a building block that reappears in each subsequent level of analysis. Thus, the principles of social action and agency developed in the study of social relationships carry over into subsequent levels. Actors continually alter the rules, agreements, policies, procedures, expectations, rituals, rewards, resources, goals, values, and ideologies which produce and reproduce social relations at and between levels. Second, the pressures and constraints imposed on a level from levels more distant in space and time are experienced, interpreted, and managed through social relationships. This is the classic definition of the sociological imagination (Mills

> 1959). Structural explanation operates *down* the levels of social structure from the world system through the domestic stratification system, through inter- and intraorganizational relations, to actors embedded in social networks, and relations. The social relationship, then, is the point where "agency" and "structure" converge. Or as Giddens (1979) would say, where the *reproduction* of social structure becomes a variant *production* of it. (Prendergast and Knottnerus 1994:6)

In this framework successive levels incorporate the capacities and pow ers of social relationships, while adding new elements, thereby creating new, more complicated forms of interdependence. More substantively the framework is described in the following terms:

> The social network is a concatenation of social relationships; its causal powers stem from the multiple and cascading *opportunities* for social interaction and exchange that connected relationships provide. An organization is an ensemble of social networks under a single canopy of *authority*; here authority relations are the key to behavioral regularities and outcomes. Interorganizational relations recycle network processes in the context of *state authority*; the new element here is the regulatory and coordinating role of the state and the professions. Overall social structure refers to the basic patterns and processes of *social stratification* in different institutional realms. Finally, the world system, the crucible of differentiation between stratification systems, is the largest unit of structural analysis. Each level assumes the existence and causal powers of the preceding levels, if not subsequent ones as well. The "scope of effects" (in terms of the nested series) is greater at subsequent levels of interdependence. (Prendergast and Knottnerus 1994:6)

To elaborate a little further, the level of social relationships is concerned with the intrarelational processes involved in the formation, maintenance, and extension of social relations and how these relationships are affected by social structures. What is involved here is a social psychology of relationships. Social networks involve a collection or assortment of connected social relationships. A loose or informal affiliation resulting from the accommodation of personal interests is at the heart of the concept. The structural quality of a network derives from the enhancement of possibilities, power, and prestige related to positions within it.

Intraorganizational relations refer to formal organizations that, drawing on Stinchcombe (1973:24), are "social arrangement[s] in which the activities of some people are systematically planned by other people (who, therefore, have authority over them) in order to achieve some special purpose." These planned activities lead actors to form distinct social relations. Planning both guides and facilitates various aspects of organizational life including task performance, delegation of responsibility and duties, coordination of activities and resources, and systems of evaluation, compensation, and discipline—features usually considered to be essential elements of what is referred to as a bureaucratic organization. Interorganizational relations focus on organizational environments that refer to environments composed of other organizations (see Warriner 1981). Here the units in patterned relations (i.e., social structures) are organizations. A new type of interdependency distinguishes this level: regulation and coordination by national and state governments, essentially through legal frameworks. Government and law heavily impact organizations and the relations between them, leading them to embrace legal-rational forms of authority and legitimizing beliefs.

Societal stratification deals with the ways in which whole societies may be structured, particularly as this relates to patterns of inequality between large-scale groups or social categories. Traditionally the study of stratification in sociology places great attention on the economic or class structure of society determined by conditions such as one's relationship to the means of production or the kinds of skills and capital existing in the marketplace and differentially possessed by actors. More broadly, societal stratification can involve structured inequalities between various kinds of groups such as political parties, elites and nonelites, racial and ethnic groups, gender groups, or occupational strata. The world system is the final level of social structure and operates at the most macro level. Here, interdependency exists between different societies, regions, and sectors of the globe. Economic, political, social, and cultural relations are extended across societies resulting in this final level of the social order. Various processes operate between societal units in this system ranging from capital accumulation and cooperation in trade and political relations to exclusion, exploitation, state de-formation, and underclass development.

This analytical scheme provides a useful framework for understanding how social rituals can operate at different levels of the social order and ritualized activities occurring in a particular setting can influence the rituals that develop in different contexts or structural levels. While employing this basic framework and its attendant assumptions about the more substantive characteristics that define the workings of each level, I will not necessarily restrict myself to these assumptions in order to include other potentially relevant conditions and processes that might be found at these different levels. In essence, though, these six levels provide a basic template for examining ritual dynamics in society.

More precisely, ritualization processes, I argue, can move from macro to micro and micro to macro levels or can occur within the same level. Conceptually, these processes can involve the inter-level and intra-level transmission or spread of RSPs throughout the social order.

Inter-level transmission refers to the transmission of RSPs across different levels of structure. Examples might involve a ritualized practice spreading from the global/world-system level to the societal level in a particular country, or the transmission of RSPs from the relationships among actors in small groups in an organization to a wider array of social networks or the organization itself, that is, the intraorganizational level. Ritualized activities can move from macro levels to more micro levels or flow in the opposite direction from micro levels to macro levels of society. Moreover, the transmission of ritualized behaviors can hypothetically involve either structural reproduction or structural transformation. The spread of RSPs could result in either the reproduction or replication of ritualized practices and patterns of behavior or the alteration of ritualized actions and social arrangements.

Intra-level transmission refers to the spread of RSPs within the same structural level of the social order. This involves a "lateral" type of ritual transmission framed within the context of the levels scheme. Here, ritual transmission occurs within the same level of analysis whatever structural level that might involve, for example, the spread of ritualized behaviors across different sectors or large parts of society or the transmission of RSPs between different informal groups and the relationships among actors within each of these groups. Here too, the diffusion of ritualized activities could theoretically involve either the reproduction or transformation of ritualized behaviors and arrangements.

And again, the six structural levels of the social order highlighted in this scheme involve social relationships/social interaction, social networks, intraorganizational relations, interorganizational relations, societal (stratification), and the global/world system.

The levels of social order and ritual dynamics discussed above are depicted in Figure 2.1. The vertical arrow, pointing down and up, represents respectively the inter-level transmission of RSPs from macro to micro levels and micro to more macro levels. The horizontal arrow pointing left and right denotes the intra-level transmission of RSPs across the same structural level.

Figure 2.1 Model for Levels of Social Order and Ritual Dynamics

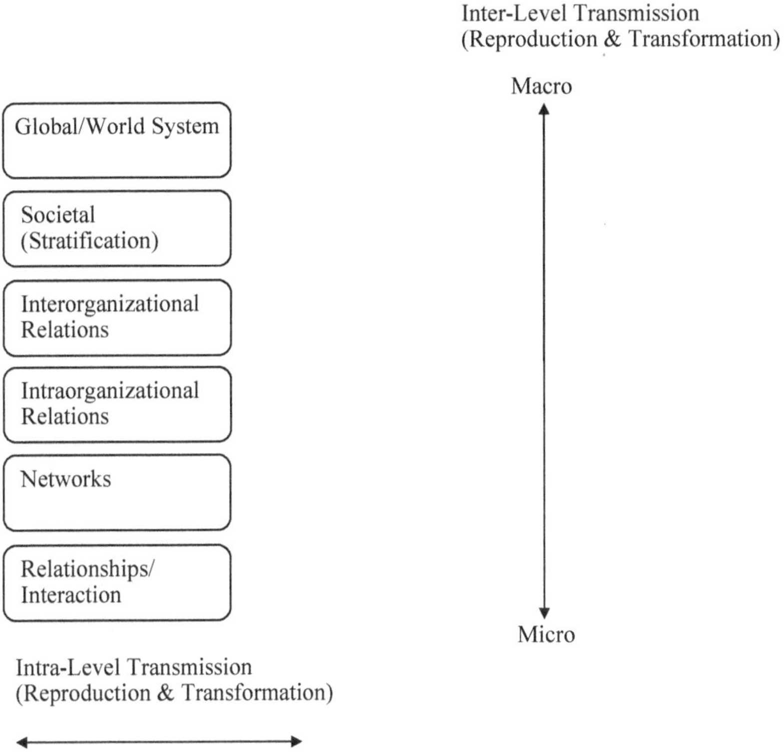

This model depicting levels of social and ritual dynamics helps us to conceptualize more clearly how rituals may be transmitted throughout society and have different consequences. At the same time this discussion raises a number of critical questions about ritual transmission including the following: What are the ways—both empirically and analytically—that RSPs operate across levels and within levels? How do both reproduction and transformation involving the spread of rituals specifically occur? Or to put it a little differently, is it possible to predict to varying degrees the operation and effect of RSPs across and within levels?

For example, in the presentation of the theory discussed earlier in this chapter and a number of studies that have been conducted in recent years, the focus is on how highly ranked RSPs in a wider environment can lead to similar ritualized behaviors and arrangements—structural reproduction—in embedded groups. This would be a case of inter-level transmission going from a macro to a micro level. For instance, a slave society in a slave plantation being influenced by the social system of the plantation would involve the transmission of ritualized practices from the intraorganizational level (the plantation) to the level of social relationships and interaction among the slaves themselves (the slave society).

In the earlier presentation of the theory, one type of transformative ritualization is discussed in which different ritualized practices in a wider social milieu can lead to a change in the rituals and structure that emerge within an embedded group. This too would be an example of inter-level transmission (albeit of a transformative nature) going from a macro to a more micro level. Of course, ritual change can occur in other ways. For instance, a political advocacy group, a civic group, or a celebrity might be able to initiate changes throughout an entire society by promoting new ritualized practices among the members of that society. Such an occurrence could include inter-level transmission moving from a micro level involving individual actors or an organization up to the macro level of society as a whole; or, agencies within a formal organization such as a government bureaucracy might initiate changes in the operating procedures of that organization, resulting in the development of new ritualized behaviors among many or all of its employees. Since the transformation in ritual practices in the organization emerges from subunits and individuals within the organization itself and is grounded in the authority system of its bureaucracy this could be conceptualized as an example of intra-level transmission.

In that regard, the levels model also enables us to conceptualize different forms of intra-level ritual transmission. It might seem more likely or apparent that the spread or transmission of RSPs within any structural level would involve the reproduction of ritual practices, but, as the last example shows, ritual change can also occur within levels. Whether such transmission involves the replication or modification of ritual enactments a number of examples can be briefly mentioned.

At the level of relationships and social interaction the spread of rituals could involve small informal groups influencing the ritual dynamics and social arrangements of other loosely formed groups, whether those involve peers, friends, neighborhood groups, social cliques in a school or workplace, and so on. As for the second level, RSPs might spread between wider networks within a community such as one neighborhood's network affecting another neighborhood's social network in a city or town or a social network grounded in a particular ethnic/racial group influencing another ethnic/racially based network in the same community or another city.

At the intraorganizational level ritualized activities might be transmitted from units within an organization such as a hospital, the military, a university, a corporation, or a social welfare agency to other units, divisions, offices, or departments in that same establishment. In regard to the interorganizational level RSPs could be transmitted from one organization or institution to another. For instance, ritualized behaviors involving consumerism might spread over time from the market or economic sector of contemporary society to other institutions such as churches, shaping their organizational structure and the religious practices of their members. Or it is possible that in the modern world we live in consumer- and market-oriented ritualized behaviors have spread into other organizational arenas such as higher education or both professional and nonprofessional sports.

At the societal level various social units and population groups usually implicated in the structuring of societal inequality or stratification may transmit rituals to other groups in society. State political systems, social classes, occupational groups, educational groups, corporate elites, and status groups (involving different kinds of status categories such as racial and/or ethnic distinctions or religious distinctions) could be the source of ritualized practices and distinctions that spread throughout society influencing, for example, other social classes, large-scale racial/ethnic group hierarchies, or workers. Another example would involve political leaders and groups in modern state systems or centralized political structures being the source of RSPs that are transmitted to the citizens of a country, sometimes in an authoritarian and manipulative manner and sometimes in what some would consider more legitimate, institutionalized ways.

Finally, at the global/world-system level societies and global regions or sectors may transmit ritual behaviors to other societies and sectors in the world. Examples would involve ritualized American clothing styles and fashion, sporting and recreational practices, religious activities, or linguistic expressions spreading to other areas of the world such as European societies; activities common to Western societies such as certain kinds of legal practices, political orientations, and ritualized managerial styles and business practices spreading to non-Western countries; or in ancient history the transmission of ritualized Chinese practices involving death and burial or styles of dining to other countries such as Japan or Korea.

Of course we should remember that intra-level and inter-level transmission of rituals can be much more complicated than suggested by the description of these concepts and the examples provided. In reality both can occur in conjunction with each other, sometimes simultaneously. For instance, the intra-level spread of activities such as recreational practices, consumer rituals, or ritualized managerial styles from one society to another may be followed in the second society by the inter-level transmission of these ritualized behaviors from these most macro levels to more micro levels. As these ritual practices flow into the society and to different sectors, such as certain classes, status groups, religious institutions, or occupational groups, they may be further reproduced in different types of organizations, then at the network level in various assortments of extended associations, and ultimately at the most interpersonal level of relationships involving individual's family lives, dating behaviors, relations with friends, interaction with neighbors, or contacts with relatives.

Social Conditions for Ritual Transmission

The discussion of ritual transmission and the complexity surrounding such ritual dynamics raise many kinds of questions. In that regard I would like to direct our attention to one final topic that warrants further consideration. This issue concerns the question of what are the social conditions and mechanisms in society that may help to transmit RSPs, whether that involves inter-level or intra-level transmission and the replication or modification of ritual practices. In other words, what are the social factors, conduits, and developments in society that play key roles in the spread of ritual enactments? While a comprehensive answer, both analytically and empirically, will not be offered here, several factors can be identified that clearly indicate how important this question is and the need for much more attention to such issues in future work.

To begin with, social structural arrangements can be incredibly important for the transmission of rituals. The prior discussion of SRT and the next chapter all focus on this social condition emphasizing how when highly ranked RSPs are presented to groups structurally embedded in large social environments this can lead to the reproduction of social patterns and ritual enactments.

Another important factor involves the role of social and political power and how groups may in calculated and deliberate ways promote the spread of rituals. Political systems and governments may actively implement ritual practices in order to achieve certain ends in a society. And groups other than political systems may put into effect RSPs in a social environment such as a community or a society. Examples of such groups could range from religious groups or businesses to ethnic groups. Furthermore, the reasons for and the ways in which groups might carry out ritual practices may vary. One group, for instance, may instigate rituals in order to affirm their social identity while another may promote certain practices in order to highlight some concern they have and to increase the group's visibility in the wider community.

The role of organized groups or social organizations in ritual dynamics and transmission is also extremely important. Here I should point out that while intraorganizational relations make up one of the levels in the six-level model, organizations are so ubiquitous and can play such a pivotal role in ritual dynamics they also deserve special attention as a social factor that can facilitate ritual transmission. Moreover, it should be appreciated that while organized groups can involve the kinds of social organizations referred to in the levels model (i.e., large formal entities marked by extensive planning, legal-rational authority, usually a number of extensive bureaucratic features, and a significantly wide scope of effects and many nested ties), they can sometimes be much smaller and have, relatively speaking, fewer nested ties and a narrower scope of effects. For instance, local organizations are often found in ethnic urban communities in the United States that promote various cultural and ritual practices such as special dinners, festivities, religious events, and recreational activities. They are organized to varying degrees, involve some planning and coordination of activities and materials, and often are based on a legitimated system of rules and procedures, yet they may be quite small (e.g., having only twenty, fifty, or one hundred members), and their effects may be limited to just the local ethnic community. For this reason, in discussing social factors affecting ritual dynamics I am concerned with all types of organized groups or organizations, not just those groups that exhibit the qualities associated with social organizations in the levels scheme.

How organized groups are important though can vary, taking different forms and having different effects. On the one hand, organizations can facilitate the development and transmission of RSPs throughout the organization itself. Interventions or formal programs developed by those in authority in an organization could create new RSPs that alter the behaviors of those working or residing in that setting such as staff in a high school instituting a program aimed at changing the ritualized behaviors of high-risk youth engaged in drug taking, criminal behavior, or unwanted pregnancies. Here, the internal authority system of the organization can play a major role in promoting rituals that affect the internal dynamics of the collectivity.

On the other hand, organizations or units within an organization may transmit and reinforce RSPs that influence the surrounding social environment. Examples would range from ethnic-based organizations such as cultural groups to political or religious organizations, civic groups, or sports organizations encouraging ritualized activities that impact an ethnic community, the wider population in an urban area, or an entire society. What's more, various circumstances or conditions specifically related to organizations may facilitate or impede the spread of RSPs in a social milieu.

To name a few, actors being circulated or rotated in and out of special groups or periodically attending special group meetings such as motivational meetings, retreats, or certain kinds of training sessions within a business organization could create or reinforce RSPs that are carried back into the wider social setting. For instance, attending special training and motivational sessions in a large corporation could promote distinct ritualized behaviors and attitudes that then permeate the wider culture of the organization.

Another factor could involve actors' placement or voluntary entry into organizations for extended periods of time in which they are inculcated with new RSPs (derived from either staff or fellow residents), which is then followed by their reentry into the wider society. To illustrate, after leaving institutions such as prisons, reformatories, or the military, where people are often exposed to various kinds of RSPs, individuals may continue to engage in these ritualized practices in the wider community sometimes to the benefit of and sometimes to the detriment of themselves and others.

Other kinds of conditions can also affect an organization's ability to successfully communicate with people in the wider society, thereby influencing the degree to which certain kinds of ritualized practices promoted by the organization are adapted in the larger environment. Here we are concerned with factors that influence the channels of communication that exist between an organization and the wider society and the simple fact that the more people can be directly reached in the wider society, the more likely RSPs will spread. On the other hand, the fewer people that can be directly reached, the less likely RSPs will be transmitted out of the group. Conditions affecting degrees of communication could involve the number and nature of links the organization has to other groups and individuals (e.g., many versus few connections with different civic groups), access or lack of access to modes of communication (e.g., newspapers, radio, journals, television, internet websites, newsletters, spokespersons), and the kinds of formal and informal ties the organization has with others in the wider environment (e.g., close ties with persons in the community possessing large degrees of influence and social contacts).

Various internal organizational factors can also enable an organization to transmit RSPs to the wider society. Such conditions would include the kind of leadership in the organization, resources possessed by the group, the nature of its authority structure, and whether this structure facilitates the organization's ability to promote RSPs in local communities and possibly at the larger national level. Clearly, organizations and how they are structured can strongly influence, albeit in different ways, the diffusion of ritual practices within both themselves and society at large.

In addition to factors such as structural arrangements, social power, and organizational dynamics, a number of other factors can impact ritual transmission. A cursory examination of the modern world reveals various processes and conditions within societies that can influence the spread of rituals. Very briefly, examples would include, but would not be limited to, written and electronic communications and the internet; artistic mediums and literature; mass media; speakers and presenters from one societal sector addressing people in another institution or sector (e.g., experts in marketing/business making presentations to religious groups); and legal regulations and procedures emanating from government applied to other sectors and organizations in society (e.g., state or federal laws applied to financial institutions or health care facilities).

So too, contacts between different societies can be a very important conduit for the flow of rituals between cultures. Such contacts can occur in many different ways and in both modern and premodern societies. Illustrations of these kinds of social processes facilitating the spread of rituals would include individuals traveling to and from other countries (e.g., emissaries, exchange students, vacationers); invasions and the colonization of societies; migration and immigration; religious missionaries and proselytizers operating in countries where the religion was not originally present; exploration and expeditions involving contact with new groups and cultures; and international organizations and agencies operating in countries in different parts of the world (e.g., NGOs, agencies with the United Nations, international advocacy groups, international legal organizations, international relief agencies).

Finally, I would highlight two other types of social occurrences that can profoundly influence ritual dynamics. Social movements have oftentimes been a source for the dissemination and creation of ritual practices. Political, religious, ethnic, protest, and other kinds of social movements very often incorporate and help spread different kinds of rituals throughout societies (e.g., the French Revolution or the Cargo Cults of New Guinea). So too, special collective ritual events including festivals, political rallies, and religious gatherings can provide important occasions for people to assemble and engage in common activities that celebrate their commitment to some group, cause, and/or ideal. Such collective events and social movements can both reinforce preexisting RSPs and serve as a conduit for the spread of new ritualized practices throughout the different parts and levels of the social order.

All of these social mechanisms and conditions are relevant to our understanding and study of ritual transmission and ritual dynamics more generally. This limited discussion, however, is not meant to provide an encyclopedic listing or comprehensive analysis of such factors. It is meant to sensitize us to their importance and how in their many different forms they contribute to the ubiquitous presence of rituals and their impact on our lives. Or, to put it a little bit differently, this brief discussion was meant to further reinforce the value and need for research, grounded in theory, of ritual in human behavior.

In that regard, the analysis of any particular dimension of ritual dynamics requires substantial effort, focused study, and ideally a cumulative or growing body of knowledge. That is why in the following chapters I will discuss several substantive issues employing the theoretical concepts we have introduced here, including the levels scheme. Some of this work will also build on the ideas discussed in this chapter and present new concepts. We first turn to research examining the reproduction of ritual and structure in groups across different levels of analysis. After that, our attention will turn to how rituals may be used by groups to achieve certain desired goals. And then we will examine a new topic that goes to the heart of SRT, that is, the crucial role that rituals play in our lives as revealed through an examination of extreme disruptions to the ritualized practices that make up our taken-for-granted daily lives.

Chapter 3

Structural Reproduction in the World and Throughout History

The formulation dealing with structural reproduction described in the previous chapter takes the form of an abstract, generalizable theory that can be applied to an array of cases for the purpose of explanation. That is why I, and a number of different researchers, have in recent years conducted studies examining the social dynamics occurring within embedded groups in very different settings. Underlying all of these investigations is a commitment to multimethod research. A variety of methodological strategies can be used to investigate a theory such as this. This research can take various forms ranging from laboratory experiments, surveys and interviews, or ethnographic case studies to historical investigations and the analysis of literary documents (for the argument that formal theory can use historical and qualitative evidence, see Collins 1984).

A wide variety of studies and settings demonstrate that structural reproduction occurs in social life. It occurs in the modern world and in the past, including ancient societies. And it occurs in quite different societies and cultural groups. Indeed, evidence suggests that ritual dynamics such as the type discussed here are found throughout the world. These findings have been obtained in investigations of a variety of settings and historical periods, including experimental groups in the United States, youth groups in nineteenth-century France, slave societies in American slave plantations, and youth trained in an ancient social system.

These studies show how powerful these ritual processes are because people will reproduce behaviors and ritualized arrangements based on simple exposure to such practices for limited periods of time, when there is no reward or benefit derived from doing so, even while they may be verbally expressing opposition to these practices and when engaging in such behaviors may lead to undesirable consequences for the group. The need to draw on the ritualized actions, symbols, and resources from the social environment surrounding us to construct—or, to be more precise, reconstruct—our immediate social worlds is both powerful and seemingly unavoidable. As previously noted, Simmel recognized this basic tendency many years ago when he observed how secret societies often end up resembling the very society they are opposed to and that they seek to remove themselves from.

Of course, it should be remembered that a theory such as the one presented here should be able to stand the test of falsifiability. A theory should be capable of tests which would or would not provide evidence which supports (i.e., confirms) its ideas and predictions. Tests of a theory along with other research which provides exemplifications or illustrations of its arguments all contribute to our confidence in the theory's ability to explain social behavior.

Research on Structural Reproduction

To begin with, an experimental test examining the reproduction of social structures in task groups among American college students has provided strong support for the theory (Sell, Knottnerus, Ellison, and Mundt 2000). Utilizing SRT and expectation states theory, this experimental study examined forty-four small task groups under three different conditions—a baseline condition in which group leaders observed no other task groups, a condition in which leaders first observed two groups characterized by hierarchical rituals and authoritarian leaders, and a condition in which leaders observed two groups characterized by lesser degrees of hierarchical ritual and procedures. We predicted that in this study the leaders would transform their own small talk groups in the image of those they saw. This would happen even when the observed groups were not involved in tasks similar to the tasks assigned to the leaders and the group, there were no incentives for engaging in particular ritualized behaviors and procedures, the leaders of the group were the only persons to observe other groups, and their exposure to other groups was of a very short duration.

Essentially, the findings of the study supported these predictions and arguments. The leaders of groups transformed their groups into ones that mirrored the highly ranked ritualized social practices to which they had been exposed. This occurred because the high rank of the RSPs within observed groups changed leaders' actions and perceptions so that the new groups were created in the image of those they were exposed to.

Given these findings, this is a good time to mention that this formula tion, and SRT more generally, relates to the arguments of several other perspectives including social cognitive theory, which is most closely associated with the work of Albert Bandura. Social cognitive theory (which normally uses the experimental method) stresses the significance of observation and modeling. Although reinforcement of the model is a key part of the theory, behaviors can still be modeled if there are no consequences to the model. For Bandura modeling is not simply imitation, but instead involves "symbolic transformation" (1986:56). This perspective is essentially a psychological theory that explicitly focuses on the cognitive abilities of people. It does not deal with structure or social interaction. While consistent in certain ways with SRT, it does not address the specific concerns of this approach—RSPs and the role they play in social life, group dynamics, and social structure. Consequently, while the central principles of the two approaches are similar (social cognitive theory being broader), SRT specifically focuses on the conditions under which RSPs may transfer to other settings.

SRT also relates to the mimesis principle in the organizational new institutionalism perspective. This concept stresses that organizational arrangements come to reflect the rules and form of other, at times larger, cultural and social structures, including the state (Meyer and Hannan 1979; Meyer and Rowan 1977). Analysts such as DiMaggio and Powell (1983) address mimetic processes and suggest that organizations model other organizations as a reaction to uncertainty (for a useful discussion of the new institutionalism approach, see Powell and DiMaggio 1991). Although the modeling portion of this perspective resembles parts of SRT, the latter is much more explicit about people's role in the process, the kinds of situations under examination (i.e., the occurrence of RSPs in specific types of settings), and the dynamics and conditions through which reproduction occurs. In contrast to the organizational perspective,

SRT's focus is on the agents and processes by which certain kinds of isomorphism arise (for a discussion about the differences and similarities between SRT and the new institutionalism approach, particularly as it relates to the "institutional logics" of organizations emphasized by Friedland and Alford 1991, see Troyer and Silver 1999).

With that in mind we should appreciate that research strategies other than the experimental method can provide important evidence and tests of a perspective such as SRT. An extended discussion of all the research that has been conducted dealing with structural reproduction using the theory is not, however, possible. For that reason, I will describe more fully one historical comparative investigation and the evidence used in this study, which is quite different from that used in an experiment.

Historical comparative research provides another extremely productive strategy for generating evidence that allows one to evaluate a theory. This type of research can, however, take different forms. For instance, it is possible to study the same case at different points in time within the same society. Changes in the larger social environment and in the embedded group at different periods would provide a basis for evaluating the theory's capacity to explain various social outcomes. It is also possible that a case could be studied during the same time period but within different conditions or societies. A judgment could then be made about whether differences in environments and embedded groups were consistent with the theory's arguments.

Nineteenth-Century Elite French Schools—A Comparative Analysis

The latter approach is what was used in a recent study of female and male youth societies in nineteenth-century French schools (Knottnerus and Van de Poel-Knottnerus 1999; Van de Poel-Knottnerus and Knottnerus 2002). Historical comparative research examined male and female youth groups embedded within two quite different elite educational institutions during the same time period and in the same society. By using this approach we were able to determine whether the arguments of the theory were consistent with and explained the social outcomes revealed in the study, that is, if these youth societies mirrored or reproduced the respective institutional systems they were embedded in.

This particular subject is particularly conducive for a historical comparative study since elite schools for girls and boys were quite different in French society in the nineteenth century. Through most of this century, the daughters of the privileged classes were educated in boarding schools, in particular, in Catholic boarding schools. Many of these schools were in convents. Here, religious and moral instruction played a central role in girls' education. It was only in the later part of the nineteenth century that a state-sponsored secondary school system was developed for young women.

A very different situation existed for boys. After the French Revolution a movement emerged to make secular, rather than religious, education the dominant mode of training for male youth. This secondary school system—state lycées and municipal colleges—was created in the beginning of the nineteenth century. Initially the goal of these schools was to create military leaders and officials to administer French territories, but these goals were quickly replaced by the more general desire to prepare young males for various governmental, military, professional, and other high-ranking roles in French bourgeois society. Many of these lycées were built in more rural areas of the country. Wherever they were located, however, they usually resembled military barracks with high walls and imposing buildings, with life in them assuming a cloistered, permanent quality.

What resulted were two distinct educational institutions for the daughters and sons of the elite classes. Youth were placed into these two types of schools and brought together to study and live, normally as boarders and to a lesser extent as day students; this made possible the development of their own unique social worlds, which were often partially if not totally hidden from the eyes of the staff who ran these institutions.

The study of social historical developments such as these is difficult, however, because although some data exists concerning the organizational characteristics of these educational institutions, information about the informal and in many cases secret social activities and experiences of youth is quite difficult to find. Fortunately, information relevant to this investigation is available in various kinds of literary texts. In the nineteenth and twentieth centuries a tradition of French literature in the realist school emerged in which authors wrote about their own and others' experiences in these religious boarding schools and lycées. Using these literary sources and, to a lesser extent, historical records, we can describe the nature of these social worlds.

More precisely, to carry out this study, we employed a particular technique that enabled us to describe the social dynamics operating in these very difficult to study youth societies. This research strategy, which uses literary sources, provides a crucial body of evidence about the nature of these youth groups located within the social environment of these schools. Indeed, without this evidence it would be difficult, if not impossible, to conduct this historical comparative investigation. I will first briefly describe how this strategy should be carried out and then discuss some of the findings from the study.

We refer to this mode of investigation using literary documents for the purposes of social description and analysis as a "literary ethnography." This methodological strategy carefully follows a set of guidelines that allow for the rigorous and systematic use of literary sources in research. By following these guidelines researchers can have greater confidence that literary texts can be used to understand and explain social occurrences. While this is not the place to discuss how other scholars have viewed the use of literary narratives for research, I would point out that a number of researchers within the area of the sociology of literature have emphasized how valuable literary documents can be. For instance, Lowenthal (1961; 1986) has argued that through a selective, creative process, a writer develops a portrayal of individuals' approaches to and perceptions of social reality involving aspects such as class dynamics, emotional relations, work, intimate relations, and close associations with others. When literary documents are used for the purpose of analysis, a representation can be produced of actors' participation in groups and their wider social milieu in different historical periods.

We argue that a literary ethnography grounded in a focused examination of literary narratives, such as novels, short stories, biographies, autobiographies, fictionalized autobiographies, memoirs, plays, or diaries, can provide an accurate description of a social environment and the key developments and social relations in this setting. We also stress that to do this, several procedures must be followed if researchers are to have the greatest confidence that this kind of evidence creates a reliable basis for analyzing social events. Moreover, other requirements must be met. These include the necessity of not limiting this type of research to only one or a few literary documents. Instead, multiple—meaning many—literary sources should be studied. Whenever possible, other information obtained from alternative methodological approaches should also be collected (e.g., archival/historical evidence and analysis, interviews, and evidence derived from other sociological methods). Lastly, the literary narratives should belong to what is referred to as the realist tradition.

Let me in a few words describe the procedures or set of six guidelines that define a literary ethnography—an approach, I would add, that can involve either an interpretative process in which a researcher engages in an intensive reading of documents (similar in certain ways to research in qualitative sociology) or quantitative evidence or even statistical procedures (e.g., counting how often literary narratives refer to certain occurrences or themes). With that said, the present study rests on interpretative procedures, and the guidelines are essentially framed within this context.

It is first necessary that the *scope of literary sources* to be studied is defined. One must determine what kinds of documents are available in terms of the issue that interests the analyst. To develop clear boundaries about the range of documents subject to inspection means that the researcher must have a clearly developed idea about the topics that she or he wants to examine and the theoretical perspectives that are relevant to these concerns. One should have a clear idea about what kind of group, institution, society, aspect of people's lives, and/or time period is of interest: for example, the experiences of teachers in lycées in the nineteenth century or the social world of plantation slaves in antebellum America. And again, the most judicious strategy requires the researcher to confine his or her investigation to a tradition of literature that appeared in the late eighteenth century and continues to the present known as the realist style (among the many writers associated with this tradition are Zola, Balzac, Stendhal, Hemingway, Faulkner, Dreiser, James, and so on).

The next step, *reading and interpretation of works,* involves how to read the literary narratives. This means that a researcher must engage in a focused concentration and reading of all or significant portions of the works identified in the first step. Some or all of the documents will almost inevitably have to be read several times. A number of considerations come into play here. For instance, we suggest that the collection of texts should not be restricted to only well-known writers and those defined by literary audiences as exceptional. The goal of the research process is to examine the widest array of texts, not just writers of a certain status. Moreover, it is essential that an extensive collection of texts be read. While impossible to provide an absolute number of sources, at least several dozen if not a hundred or more documents probably should be read. For instance, in the study of youth societies in French elite schools, approximately 150 literary sources were examined.

The *identification of textual themes* involves the discovery of different themes that refer to people's experiences and the social setting. Here the researcher is first recognizing that certain thematic elements are appearing in different sources—that is, this is the first indication that certain thematic elements appear in different texts. As themes dealing with, for example, social relations between group members or rewarding or negative events in people's lives are identified multiple readings of documents will probably be necessary.

After certain themes have been identified the *classification of thematic elements* occurs in which a classification scheme is developed that represents these themes. Now the researcher formally categorizes thematic elements that are referred to continuously. The categorization system demonstrates that one has discovered clear motifs in many texts, that is, certain elements that appear frequently in the various literary sources. This categorization scheme can be considered to be a collection of empirical generalizations about social conditions, occurrences, social and personal experiences, and so forth that are repeatedly addressed in the collection of literary documents. These thematic generalizations generate a descriptive account of people's social conditions and experiences. These, together with the next step involving the analysis of these materials, provide an overall representation or depiction of the phenomena being studied.

The *development of analytical constructs* refers to abstract ideas or generalizations concerning the qualities of the subject matter being studied, that is, the thematic generalizations categorized in the previous stage. Analytical constructs contribute to our understanding of the phenomena because their qualities or features are interpreted in terms of abstract formulations or principles. While our focus here is on sociological and social psychological formulations, these analytical constructs could draw on abstract concepts and theories in any of the social/behavioral sciences. Such constructs enhance our understanding of and ability to explain phenomena such as the nature of the social relations between actors in groups, the kinds of social structures that distinguish these groups, and how and why these social relations and the way they are organized emerge and are perpetuated among group members.

Finally, a *contextual confirmation* must be conducted by rereading the collection of materials to ascertain if the narratives correspond in an accurate and meaningful way with the classification scheme and analytical constructs. The researcher needs to reexamine (perhaps numerous times) the literary texts to be certain that the identified motifs and abstract concepts are correct. By confirming that commentary in the literary sources is consistent with the thematic categories and analytical constructs our confidence in the findings of the study is further strengthened. The result of this last step and the entire process is the development of a composite or ethnographic portrait of the subject matter, such as the personal and social world of a group of actors.

In sum, when carried out in a careful, rigorous, and judicious manner a literary ethnography can provide us with another powerful tool for conducting social research and better describing and understanding social behavior, group dynamics, and larger groups in society.

Having described this research strategy let us examine some of the findings obtained from our historical comparative study of boys' and girls' experiences in the elite schools of France. We will begin by summarizing some of the most important findings concerning the educational institutions for boys and the kinds of social worlds that existed among those students. Then the same will be done for the girls' schools and informal social world. Following that a number of literary narratives will be presented that substantiate these points.

In nineteenth-century French lycées we find that male youth recreated many of the ritualized behaviors and relations they were exposed to in their informal, oftentimes secret, embedded world. More precisely, this elite school system was an extreme type of total institution, which, from the perspective of the theory, contained a distinct set of highly ranked RSPs and relational features. These dominant ritualized activities engaged in by staff members such as principals, supervisors, and teachers, which male youth were exposed to in this institutional environment, included a highly formalized, quite inflexible, and complex hierarchical authority system. Rigid, hierarchical relations characterized the interaction between staff and students.

Staff possessed a great deal of authority while students essentially had no authority. So too, punishments and extreme discipline were central to the ritualized behaviors of staff as they sought to control all aspects of these youth's lives. Relations between students and staff were also marked by high degrees of social distance and formality. Boys were usually dealt with in a demanding if not oppressive manner nearly all of the time. All of these ritualized practices focused on and expressed symbolic themes that emphasized the harsh use of social power, rigid distinctions, the stringent use of punishments and discipline, high degrees of impersonality, and authoritarian modes of behavior.

The personal and social experiences of youth within these educational institutions also exhibited very distinct qualities as revealed by an examination of literary sources. Briefly stated, *entry* into the institutional milieu of the school involved a major transition in the lives of youth and often included *abusive rites of passage* (italics here and in the following two paragraphs indicate key findings from our research). Some boys also experienced *feelings of loneliness.* We find, as might be expected, that a *need for acceptance* existed among some youth. Youth were also regularly subjected to various *punishing or aversive experiences.* Staff, including teachers, exercised extreme control and a watchful eye over students while placing a great stress on schoolwork both in and outside of class. It is revealing that one finds hardly any discussion of rewarding experiences in all the texts. *Classism* or *elite and non-elite distinctions* were also quite common among students and, for that matter, the entire institutional environment.

Perhaps most importantly, an *enduring* and *pervasive* feature of this youth world dealt with *group formation,* that is, the creation and perpetuation of informal groups of varying size among students. While different kinds of leaders were present in these groups, *authoritarian leaders* are most often mentioned. Many of these youth groups exhibited *rigid hierarchical social structures* or arrangements, for example, groups and gangs with very authoritarian leaders sometimes literally referred to as chiefs, boys who were clearly identified as next or second in command, and explicit pecking orders among the members.

And, *coercive* and *oppositional orientations* were evident throughout this world of youth and in many of the specific groups that boys created as seen, for instance, in their aggressive treatment of and opposition to other groups from their own or other schools, the ridiculing or mistreatment of certain individuals, extreme bullying, and scapegoating. In sum, the social world of these male youth was characterized by an emphasis on *power in social relations.*

What we find then is that while boys repeatedly expressed their opposition to and dislike for the strict regime they were exposed to in nineteenth-century French lycées, they ultimately reproduced these authoritarian, hierarchical arrangements within their own social world. The social dynamics of these youth were characterized by an emphasis on power, coercion, aggressive relations, a lack of affection, and inflexible and lasting hierarchical social structures. What resulted in their informal, hidden social world was an even more exaggerated version of the social environment they were embedded in. Or, to put it in the language of SRT, these ritual dynamics led to isomorphic structural arrangements between the social world of youth and the educational environment they were embedded in.

The elite schools for girls also exhibited the features of a total institution; however, the degree to which they were present varied. While some features such as levels of surveillance and staff consensus were present to a large degree, others such as hierarchical authority systems and social distance were less so. Being smaller in size these schools and their authority systems were not especially complex and interaction between students and staff exhibited varying degrees of closeness. While punishments (i.e., mortification practices) were used, they were also accompanied by rewards. And aversive practices common to the boys' schools such as verbal sarcasm and ridicule by staff were far less prominent in the girls' institutions. These religious schools were clearly a less extreme version of a total institution.

From the perspective of SRT, the social milieu of girls' schools possessed several highly ranked RSPs and arrangements. They were prominent or salient practices in the daily lives of everyone in these schools. Moreover, staff engaged in these ritualized activities with youth and with one another a great deal, that is, there was a high repetitiveness. These practices were also homologous because they exhibited the same symbolic themes and forms in different domains of interaction, for example, church, dining areas, classrooms, play areas, and dormitories. And various resources ranging from the physical setting of the schools to the mental and social skills of staff and students facilitated the enactment of these ritualized behaviors.

These dominant ritualized practices involved a moderately simple authority system and an array of social relations that were to differing degrees strict in nature but were also marked by affection, sincerity, and intimacy. While staff disciplined students and were firm in their exercise of authority their daily interaction with youth also involved enjoyable and rewarding experiences, consideration, kindness, and play. Moreover, although social distance and a sense of decorum influenced some of the interaction between staff and girls, relations between the two were often marked by openness, familiarity, and sometimes a very real closeness. Additionally, many of these daily RSPs directed toward students involved religious practices and values, morally desirable behaviors, and qualities associated with their (culturally defined) female gender, such as music, sewing, simple reading and writing, and an emphasis on maternal and domestic duties.

As with male students our analysis of literary narratives revealed a number of distinct motifs or themes related to the social and personal experiences of female pupils in this educational milieu. While girls had negative expectations about entering these schools, were subjected to different sorts of punishments, and an *oppositional orientation* emerged among some youth, other themes were much more dominant (italics in this paragraph indicate key findings from our research). For instance, some students experienced feelings of contentment and exhibited a religious orientation in their attitudes and behavior. Furthermore, a large number of girls engaged in *affective and cooperative relationships* in which their interpersonal contact with fellow students and staff was *emotionally infused* and characterized by*moderate to strong feelings of attachment.* Numerous youth also exhibited a *recreational orientation,* which involved happy or lighthearted relations with other girls, that is, a *play-centered orientation* and to varying degrees *cooperative relationships.* While girls formed groups, this occurred far less than with boys. A number of these groups exhibited a recreational orientation and to a lesser extent a religious orientation and an oppositional orientation. In these groups *facilitative leaders or authority figures* are sometimes found and literary narratives indicate these groups possessed *simple social structures or social differentiation* with *limited hierarchical structures.* Lastly, as in the case of male youth a *classism* expressing elite and non-elite distinctions exists among female students.

To summarize, evidence suggests that highly ranked RSPs in both male and female institutions significantly influenced the development of schema-based action repertoires among male and female youth. What emerged in the embedded groups were comparable ritualized behaviors among students. In other words, the practices and relations that developed in the youth societies nested in their respective schools were isomorphic to the RSPs in the more encompassing educational environments. In regard to female schools, while certainly authority and discipline were sometimes strictly exercised by staff and a formality and social distance existed between youth and staff, which led to oppositional behaviors among some students, these practices are referred to less often than other kinds of actions and occurrences. Much more common among youth were affective and cooperative relationships, recreational activities and play, religious practices, feelings of contentment, simple group structures, and facilitative leaders. The emergence of group attributes such as these mirrored the organizational arrangements and frequently engaged-in behaviors of nuns and other staff, which often involved closeness, affection, enjoyable and rewarding experiences, a simple authority structure, and a stress on moral/religious themes and attributes associated with a feminine identity. Clearly very different ritual dynamics operating in these schools led to quite different outcomes from those found in boys' institutions.

The use of literary evidence for this type of historical comparative research involves very large numbers of texts and quotes, far too many to be presented in the present discussion. But I can provide a few representative examples of these narratives that illustrate the kinds of evidence used to develop the ethnographic portrait and analysis of ritual dynamics operating in these different settings. Actually a limited number of quotes is probably sufficient because of the rich, descriptive nature of literary evidence. Assuming that the selected quotes accurately reflect what multiple sources and narratives reveal and the analytical conclusions derived from an examination of them—that is, the literary ethnography—then those narratives can provide powerful and compelling imagery that illustrates some of the findings and arguments of the larger study. Several quotes that refer to the nature of the institutions boys and girls lived in will be first presented. These will be followed by various narratives describing the personal and social lives of these two groups.

The extreme nature of the total institution that boys attended is vividly captured by many authors. Take for instance the description provided by Balzac:

> Standing in the center of the town, on the little river Loir which bathes the outer walls, the college presents a vast enclosure carefully shut in, containing all the appurtenances necessary to an institution of its kind—a chapel, theatre, infirmary, bake house, gardens, and a water-course.... Distance ... prevents the parents from visiting their children frequently, and the rules forbid vacations beyond the walls. Once entered, the pupils never leave the college until their education is completed. With the exception of regular walks made beyond the gates under the guidance of the Fathers, everything is so conducted as to give the institution the advantages of conventual discipline.... The punishments formerly invented by the Company of Jesus—which were as terrifying to the moral as to the physical being—remained in all the integrity of their early promulgation. (1948:252–55)

Similar images of the isolated and imposing nature of these schools are provided by many other authors. Cesbron, Barrès, and Marthelet, respectively, offer the following revealing commentary:

> The lycée ... only showed fenced windows and black walls ... "our prison!" thought François ... pushing the heavy door which was never closed and never opened wide ... the recess yard ... how small it was!

> Dusty desert, edged with asphalt, surrounded by fences and gray buildings, with its sole tree. (Cesbron 1948:27)

> All over France, these vast lycées whose exteriors resemble military barracks and convents ... (Barrès 1897:3)

> Would I ever be able to overcome the communal and jail-like impression that the architecture of this big school inspired at first sight? Six stories of galleries, our dwelling, wouldn't you know it, on the sixth floor. A long dormitory, without any intimacy, with beds pressed against each other, and just enough space between them, to be able to slip in. At the center, the cubicle of the supervisor who didn't supervise anything, only there as a symbol, who became alive only to initiate our moves or to turn the lights on or off. We were awakened by the sound of the drum and these military rolls orchestrated our whole life. Indeed there was an old Fritz, who spent his whole life complaining and who, besides his role as "the clock," was in charge of cleaning the huge navel-shaped, evil-smelling "pagoda" located in the center of the courtyard, not far from the dining hall: a foretaste, if one dares say, of the "facilities" we would rediscover while in military service, much later. But the shower episode was the worst trial: one had to run down the six stories in order to get a fairly good place in the line that formed very quickly, before reaching the cubicles where fell on us, from the sky, water alternatively burning or freezing, until the decisive words: "stop washing" which would surprise us being totally covered with soap. (Marthelet 1993:99–100)

Equally vivid are numerous writers' accounts of the social regimen, authority structure, and overall social climate within this institutional environment. Note the following observations by Gerbod, Jammes, Cesbron, Crubellier, and Bedu.

> Authority is the school directors' primary quality. Also important is the experience of the administrator, a strong general education, firmness and skillfulness in dealing with the relationship between families and local authorities. (Gerbod 1968:30)

> The principal and the supervisor having come in a great ceremony, from classroom to classroom, to read the examinations' results to the students, I was withered by the sentence I heard: Sir you have received in mathematics a double zero. (Jammes 1971:120)

> The censeur enters the classroom, a paper in his hand. Everybody stands up; being courteous, Mr. Platrier goes down the two steps (of the ros - trum) that make him look even taller, in order not to humiliate the vice-principal who is very short.... He speaks in a low voice to Mr. Platrier who answers: No? and remains in his thoughts. Then he turns towards the classroom. His voice has two more degrees of authority than that of the most assured teacher, one more than that of the surveillants généraux [main supervisors]. That is hierarchy. (Cesbron 1948:60-61)

> The "maitre répétiteur" [teacher's assistant or pion] is the main bolt of the boarding school's machinery. He is the table partner and companion of the students; he should be their guide and friend, but ends up being their jailer. (Crubellier 1979:148)

> When it is not the French teacher who strikes a violent blow on his head with his cane in order to force him to concentrate on Hugo's *La légende des siècles,* it is the pion Tapatan, who, when catching him by surprise dreaming, strikes energetically on the table: You will copy, for tomorrow, a hundred times: "I am lazy." At the ring of the bell, it is always a relief. (Bedu 2001:32)

The central role that power, supervision, punishments, strictness, and impersonality played in the ritualized behaviors of staff are further evident in other commentary. I close this section with just a few other narratives that emphasize these points from Pagnol and Gerbod:

> The one [surveillant général] of the boarding school quarter, under his gray cap, was constantly in the hallways, like a gunboat on a colonial river, but he would also appear suddenly in the courtyard at the specific time when his presence was least desired. (Pagnol 1977:174)

> The boarders ... live at the same rhythm as the institutions subjected to a monachal and military discipline. The bell, since 1825, has replaced the drum of the Empire's Lycées, but the Napoleonic rules are still applied. Life moves on, highly regulated, between the dormitory, the dining room, the classroom and the lectures, broken by a daily walk, strictly supervised, and the "outings" on holidays. Roll calls, movements in line and orderly fashion as on a drill ground, study hall supervisors in the rear, under the watch of the vice-principal or principal, up at dawn, at five thirty during summer, early to bed, everything resembles or rather foreshadows military life. Silence ... long hours of study before and after

> the ceremonial lecture, everything also resembles a Benedictine convent. (Gerbod 1968:100)

> In these sad and leprous barracks, where air and light scarcely enter, a severe discipline is in place. In Mâcon, in 1854, the walks ... are silent and mournful parades through the city. In 1835, at the Rennes collège, "... some majestic reprimands terrorize the most shy ... extra work, confinement to the quarters, a piece of bread for dinner, sequestration win over the most rebellious ones...." The movements of students in the courtyard and the hallways are silent. It is forbidden to speak in the dining room and in the dormitory. The students' desks are searched, the administration gives reprimands.... Punishments, extra work, prison ... the periods of relaxation are rare. (Gerbod 1968:103)

In the girls' schools an atmosphere similar in some ways and quite different in others prevails. While many qualities associated with a total institution were definitely present, such as a sense of separation and isolation from the wider society and children being subjected to varying degrees of strictness, the social organization and ritualized behaviors engaged in by staff in convent schools are very different from those in boys' schools.

For instance, Sand like others comments on the isolation she experienced:

> We were cloistered in the literal sense of the word. We had vacations, but I did not, my grandmother saying she would rather not have me interrupt my education leaving me less time in the convent. (1970:868)

At the same time various narratives speak of the positive (or at least both positive and negative) aspects of life in these schools such as the following comment by Wagener:

> At age ten, our young heroin will leave this kind universe (home) to enter ... les Dames de la Providence, Fort Royal. This is 1773. She will stay there four years. And, contrary to what was said, these four years will be beneficial and formative.... The idea of the Founder ... is to prepare girls "for the duties of Religion and Society." This is explained in his "Instructions" where he emphasizes that "girls' education, of which few people understand the consequences, and of which everyone feels the inconvenience, is based on the first principles of society, because being in charge of, following the laws of nature, the first care of our youth, they are also in charge of our first feelings and our first experiences;

from this comes public morals, the happiness of families and therefore this requires detailed instructions suited to such important duties." (Wagener 1999:35)

In a somewhat related vein Malraux also provides a more vivid description of the religious tenor of life within her school:

I was sent to Sainte-Clotilde's ... it certainly was a religious school. Every morning, lessons began with prayers; two hours every week we learned the catechism and our Bible history was on just the same footing as the history of France. I took part in these lessons, and for a long while I was the only Jewess among these scores of Catholics, whose faith was never put in danger. Standing, I listened to the Paters, the Aves, and the Confiteors. Sitting, I followed the catechism in my neighbor's book, and I heard continual allusions—everyday remarks or proverbs—to what were self-evident religious truths for my companions. (Malraux 1967:5-6)

Other narratives provide us with information about the social dynam ics between staff and students along with the temperament of various staff members. Among those highlighting the positive aspects of convent school life are the accounts of, respectively, Turin, Lavergne, Malraux, Campan, and Cédoz.

These little girls imitate; if you are nervous in front of them, how do you expect them to be calm? And if you speak too much or too loudly how can they keep quiet? If, instead of giving an order in a calm and soft manner, you just shout and scold all day long without even knowing what you are saying, how can you expect the children to listen to you? (Turin 1989:183)

[Marie-Stella] is in great health, in a good temper and full of grace.... She is going to be twenty-five and doesn't even look eighteen, and, with that, a calm, an authority such that the fifty-two young ladies she gives drawing lessons to obey her, listen to her and respect her as if she were a senior. (Lavergne 1903:215)

But at Sainte-Clotilde, the unfrocked nuns treated me with great consideration; indeed, perhaps they were kinder to me than they were to the Catholics. They behaved rather as though I were second cousin to Jesus, and they told me about the joy that would reign in heaven the day of my conversion, and how I should also ensure the salvation of my relatives.

> And I was quite ready to believe them. I knew the kindness I enjoyed at school did not always obtain in the outside world. (Malraux 1967:52)

> [As a reward for doing well in different subjects she] received five "cards of contentment." (Campan 1825:147–48)

> The title of "daughter" in the boarding school of the Dames Anglaises resulted from the adoption of a student by a nun. The child would first obtain the approval of the nun of her choice, then the agreement of her parents and of Mme la Supérieure. From then on, the young adoptee was taken under the wing of the one who became her "little mother" and was then considered, by the convent, her own daughter. To become Poulette's daughter! There wasn't a child in the whole place who didn't aspire to such luck. It is because, after the advice, sermons, even remonstrances, there generally were excellent things such as cookies, candies and sweets. However, since a little mother's customers couldn't be above the number of three, all of the covetous eyes of the school would focus on Poulette, spying the moment a void would occur in order to obtain the happiness of filling it. (Cédoz 1891:343–44)

One final account also reveals how life within such schools could be emotionally gratifying. In this case the narrative focuses on the dynamics between the female students in two quite different institutions:

> While the boarding school was growing, the school for poor children established in 1862, first in the hands of the English nuns, then run by the admirable daughters of Saint-Vincent-de-Paul, was growing as well. It counted more than eighty students. The convent, as one can imagine, never paused to take interest in it . . . there are two days per year when the two schools are momentarily united: The Saint day of Mme la Supérieure and Sainte Catherine day. . . . In summer, the boarding school welcomes its guests in the park. . . . The fact is that a blending is instantly achieved, with piercing screams entering like needles in the ears. It is liberty, equality and fraternity produced here spontaneously. . . . The boarders take hold of the little school students and . . . everybody is having great fun. (Cédoz 1891:444–45)

While statements such as these provide a clear indication of what the institutional environment was like we find that even more revealing are the numerous narratives that enable us to construct a portrait of the personal and social lives of youth embedded within these schools. In texts dealing with the social world of male youth repeated accounts refer to the role of coercive and oppositional behaviors and attitudes. This emphasis on power is manifested in various ways including the ridicule or mistreatment of fellow students. Particularly evident in these accounts is the importance of groups students formed and how such social units oftentimes involved authoritarian behaviors and hierarchical arrangements. Also prominent are references to the classism pervading these schools and French society. In regard to the latter note the observation of Sylvère:

> Our class was separated into two clear-cut camps, the elite and the commoners.... The "great mind," as it should be, was over to the side of well cleaned faces, combed hair and polished leather shoes. The other side, where I belonged, was made of hirsute and unkempt faces and heavy wood clogs. (1980:199)

The ridiculing of fellow students is also an oft-noted theme. Examples of such behavior are provided by André and Gide:

> My spectacular fall [from bed] was, unfortunately, going to bring me the wrong kind of fame. From now on I would be the imbecile, the confused one, the nitwit, the clod, the always mocked, which would add tremendously to the lowering of my status. (André 1978:303)

> This stupid success in recitation [reciting poetry], and the reputation that followed of being affected provoked the hostility of my classmates.... I was mocked, beaten, hunted down.... The torture started when school ended.... I would wait for the end of the school day filled with apprehension! And as soon as I was outside, I would slip away, I would run.... Some days I would come home in a pitiful shape, my clothes in rags, full of mud, my nose bleeding, my teeth chattering. (Gide 1958:109)

More generally, comments abound in literary documents about the role of power and coercion in social interaction and authoritarian behaviors within and between the ubiquitous groups created by boys. This is clearly seen in a sampling of comments from Gide, Fournier, Cocteau, Duhamel, and André.

> Almost instantly groups were formed. A certain Léon Gheridanisol was in the center and already taking the lead. A little older than the others, and actually more advanced in his studies, with brown skin, black hair,

black eyes, he was not even particularly strong, but he had what could be called "guts." (Gide 1925:318)

Although he was still quite small, Jasmin Delouche was one of the oldest boys in the senior school. He was very jealous of Big Meaulnes while pretending to be his friend. Before our boarder arrived, Jasmin had been cock of the roost.... With one violent shove, Meaulnes sent him reeling, arms outstretched, into the middle of the classroom. Then grabbing hold of Delouche by the neck with one hand, and opening the door with the other, he tried to throw him outside. Jasmin clung to the tables and dragged his feet across the floor, his hob-nailed shoes grating on the flagstones; while Martin, having once regained his balance, moved cautiously forward, his chin thrust out aggressively. Meaulnes let go of Delouche in order to cope with this idiot, and he might well have found himself in a tricky position if the door leading to our private rooms had not opened at that moment. Monsieur Seurel appeared in the doorway.... The fight stopped immediately. (Fournier 1966:26-27)

Dargelos was the college's rooster. He appreciated the ones who braved him or assisted him. (Cocteau 1925:13)

"I wanted to die. But since I didn't succeed, I've decided to live only for amusement, like a child or a gipsy. I've abandoned everything—father, sister, home, sweetheart.... I've nothing except my companions." "They've already betrayed you," I said. "Yes," he replied angrily. "Someone called Delouche is to blame for that. He guessed that I was going to join forces with you, and demoralized my troops when I had them perfectly under control. You saw how the raid was carried out yesterday evening, how well it went! I haven't organized anything so successful since I was a child." (Fournier 1966:95)

I returned to the provincial and so very quiet rue Médéha, in which we would lead, among the piles of paving-stones, very frightful wars against the rogues of the rival school, the rue de l'Ouest. (Duhamel 1949:13)

Adding to the sorrow, there is always among groups of children a more gifted, more devious individual who has great influence on the others and does not feel any remorse in abusing it. That is how I was to fall under the power of a character, remarkable for his skill at never letting himself be caught and for his precocious gift for using his fellow men.... Leprêtre, admired and feared, soon noticed the advantages and the pleasure he got from reducing me to a state of slavery. He employed himself to this task with the greatest brutality, the same way one subdues the primitive tribes deprived of weapons. (André 1978:308)

Repeated references to such ritualized behaviors and relations between students create a vivid image of what life was like for many boys in these youth groups. We close this section with a few additional narratives from Duhamel, André, Davanture, and Bedu, which provide further support for these findings.

School scared me tremendously. I soon felt like a very small fish lost in the wild sea. Most of my schoolfellows were vigorous boys, fighters, skillful in boxing with their fists and feet, who solved all the problems of life according to the rules of the jungle. (Duhamel 1949:49-50)

... the unmerciful laws that govern children societies. (André 1978: 303)

His [Maurice Barrès] classmates nicknamed him the "crow," because he was a "little boy with black hair, grave and remote." He would experience, during recess, a feeling of "terror-stricken weakness" ... with his pants tight above the knees with a rubber band, he felt ridiculous. Above all, his classmates did not let him forget. A child who stays by himself in a courtyard where some happy louts are having fun is always an easy, tempting prey. He was kicked, insulted, mocked. (Davanture 1975:80)

At school, like in the grown-ups' world, one always meets poor creatures, laughing stocks that are the object of vulgar jokes. It is the case for Titine, tormented by Raudot because of his extreme sensitivity, his weakness and his unfortunate tendency to want to "buy" his classmates' friendship by offering them a pencil or an eraser. During a fight in which Raudot mistreats Titine ... he (Francis) acts the way Montherlant later will when protecting his friend Aragon from the attacks of the "big ones" in the schoolyard. Francis knows he will not have the upper hand and exposes himself to the fury of the boy who is older and above all much stronger. (Bedu 2001:36-37)

Turning to the world of girls, literary sources provide a very different portrait of the social rituals and relations that shape the daily lives of many youth in their schools. Of course, it is certainly the case that some references exist testifying to the harsh treatment of students by staff and girls resisting, if not opposing, various staff and fellow students. Witness the comments of Audoux, Sand, and Malraux:

> I thought to myself that he was taking away the holy oils, and I did not dare ask whether Sister Marie-Aimée were dead. I have never forgotten the blow I got from Madeleine's fist when I clung to her dress. She knocked me right over and whispered, as she ran past, "She is better." As soon as Sister Marie-Aimée was well again, Madeleine lost her arrogance, and everything went as before. (Audoux 1911:59)

> Why our sisters, so beautiful, so kindhearted, and endowed with such noble and natural manners, had put at the head of the lower form a person of repellent bearing, countenance and dress and of speech and character to match, I cannot fathom. Fat, dirty, stooped, bigoted, narrow, irascible, hard even to cruelty, suspicious, vindictive—she at once became the object of my moral and physical disgust, as she was already for all my companions. (Sand 1979:134)

> I wasn't biased when I first entered the convent. I was more inclined towards docility than revolting. One can tell you that I arrived there without any temper or sorrow; I was more than willing to yield to the general discipline. But when I saw how stupid this discipline was and how it was applied with such meanness . . . I placed my cap on my ear and resolutely regimented myself in the "camp of the devils." (Sand 1979:879)

> I had also been completely sincere when I flung books at Monique Sainte-Rose that day at Sainte-Clotilde's, and on that other occasion when I fomented a strike at school. (Malraux 1967:111)

At the same time it should be noted that some of these comments are tempered by positive and affectionate references to other staff in the school. Furthermore, such commentary is overshadowed by the frequent descriptions of close and supportive relations between staff and girls, intimate and playful interaction between students, and girls forming groups based on close ties sometimes directed by supportive leaders. Note the accounts of Stern (Marie d'Agoult), Cédoz, Grafteaux, and Audoux:

> We were in recess. The nun who presided over the games declared, while introducing me to the students, that in honor of the "newcomer" an extra fifteen minutes would be given. A noisy cheering answered her. I was welcomed to the "rondes." I was taught the games I didn't know. I was overwhelmed with questions: How old are you? In which grade are

you entering? How long are you going to stay? ... I was greeted with displays of friendship.... By the end of the recess, I had been given a nickname. (Stern 1877:155-56)

There was nothing more opposed to the popular misconceptions of the somber aspect of the convent than the sight we witnessed today. Seeing the ease and high spirit of the parents, students and nuns themselves, makes it clear that it is the usual atmosphere of the house, that it has not been put on due to the circumstances. (Cédoz 1891:452)

We did a little mathematics and some reading, and we learned our catechism. I liked my school very much. It was in a warm part of the parsonage, and the ladies from the chateau were very nice. Sometimes they gave us a little milk. The winter passed like that. When spring came and we left Avesnes for the country, Papa, like the other heads of families that "migrated," contacted the parochial school in our neighborhood in the Seine-Inférieure, and I continued my studies there. This went on for four years. They were the best years of my childhood. (Grafteaux 1985:14)

When evening came I saw that Sister Marie-Aimée knew what I had done, but she never said a word of blame. At recreation next day she drew me towards her, took my head in her two hands and bent towards me. She didn't say anything to me, but her eyes plunged right into my face. I felt as though I were wrapped up in her eyes. I felt as though a soft warmth was all around me, and I felt comfortable. She gave me a long kiss on the forehead, then smiled at me and said, "There, you are my beautiful white lily." I thought her so beautiful, and her eyes shone so with several colours in them, that I said to her, "And you, too, mother, you are a lovely flower." She said in an off-hand way, "Yes; but I don't count among the lilies now." Then she asked almost roughly, "Don't you love Ismérie anymore?" "Yes, mother." "Really. Then what about Colette?" "I love Colette too." "Oh, you love everybody!" she said. (Audoux 1911:47)

Other commentary more explicitly refers to the presence of groups and informal associations among girls, and how they could cooperate and coordinate their actions sometimes with the guidance of a group leader as they engaged in their shared activities. A variety of narratives describe such occurrences as seen in the following selections from Bernhardt, Audoux, Maurois, and Rivaux.

It was fête day and we had two hours for recreation. We were marching in the procession along the wall which skirts the railway on the left bank of the Seine and as we were burying my pet lizard we were chanting the "De Profundis." About twenty of my little playfellows were following me. (Bernhardt 1968:41)

All of a sudden I got the idea that the Virgin could bring this miracle about. Colette thought it was a splendid idea.... She explained to me that several girls would be necessary for the nine days' prayer, and said that we must go and purify ourselves at the communion, and that during nine days we would pray all the time, so as to get help from Our Lady in heaven. This had to be done in the greatest secrecy. It was arranged that my friend Sophie should be one of us because she was so very religious, and Colette said she would talk to some of the big girls who were good too. Two days afterwards it was all arranged. Colette was to fast during the nine days. On the tenth day, which would be a Sunday, she would go to communion as usual, leaning on her stick and the arm of one of us. Then, when she had taken the holy wafer, she would make a vow to bring up her children in the love of the Virgin, and after that she would rise up straight and sing the "Te Deum" in her beautiful voice, and we would all sing it with her. For nine days I prayed with more fervour than I had ever prayed before.... Nobody noticed the little signs which we made to one another, and the nine days of prayer passed off without anyone knowing anything about them. (Audoux 1911:49-50)

The other girls were very fond of her. At first they had thought her apathetic and silent, an expanse of "still water" ... But experience showed that she was always ready to enjoy the gaiety of others, and when things got difficult she was always dependable, and at times even heroic. She was the person to be with when it was a question of "looking for the victim" (imprisoned in some subterranean dungeon)—a favourite adventure-game at the Convent. (Maurois 1953:40)

To Jeanne's solid virtue and good qualities was united a singular attractiveness, so that her elder sisters and brother strove with their parents for her society. Cheerful and joyous in temperament, she was the soul of her class exercises and enjoyments, and such was her influence that her companions used to make her the referee of their little disputes, and it was a common saying among them: "Let us tell Jeanne about it; she will settle everything." (Rivaux 1887:69)

Finally, we conclude with two narratives, one by Bernhardt, describing a particular group she was a part of and the leadership role she played in it, and another by Cappelle, referring to the wider collective sense of attachment she felt to her school and classmates:

> Then, too, I was a favorite with my schoolfellows, who frequently did my compositions for me.... Very often during recreation a whole group of us, ten or twelve little girls, would stand round, with a cage on a bench or tree stump, and watch the wonderful work of these little creatures [spiders]. If one of my schoolfellows cut herself I used to go quickly to her, feeling very proud and important: "Come at once," I would say, "I have some fresh spider-web and I will wrap your finger in it." Provided with a little thin stick I would take the web and wrap it round the wounded finger.... I was looked upon as a little authority and was made umpire in questions that had to be decided. I used to receive orders for fashionable trousseaux, made of paper, for dolls. It was quite an easy thing for me in those days to make long ermine cloaks with fur tippets and muff, and this filled my little playfellows with admiration. I charged for my trousseaux, according to their importance, two pencils, five *tête-de-mort* nibs, or a couple of sheets of white paper. In short, I became a personality, and that sufficed for my childish pride. (Bernhardt 1968:26-27)

> The hidden tears of her arrival at school are replaced by public and collective ones at her departure.... The "I" has become "we" but the "we" is disappearing: The closing of this heavy door would end a period of our lives. Behind it, we would leave some happy years, dear friends and happy memories. What would we find on the other side? (Cappelle 1867:70)

In essence, literary evidence presents two quite different portraits of life within girls' and boys' societies in the elite schools of nineteenth-century France. In both cases the social worlds of students that emerged in these educational institutions mirrored the social environment of the schools they were embedded in (and, I would add, the ritualized practices and beliefs of the wider society in that time period about what was expected of males and females, particularly those in the elite classes).

Using two quite different research strategies—the experimental method and historical comparative analysis employing literary sources—we find that groups embedded within a social environment reproduce the highly ranked ritualized practices and relations they are exposed to within their own social spheres, resulting in isomorphic social arrangements. Different kinds of evidence concerning social groups in very different cultures and time periods reveal similar ritual dynamics and outcomes.

Additional Studies

This does not, however, exhaust the studies examining this issue. Building on research that examines structural reproduction in laboratory groups and educational institutions, additional studies have investigated reproduction occurring in other institutional and organizational settings, in an entire society, internationally, and among ethnic groups where ritualized practices from quite different cultures are perpetuated. The latter studies focus on very different kinds of ritualized practices engaged in by people in a wide range of social milieus. Let me in a few words describe these four investigations of quite dissimilar groups.

First, research (Knottnerus 1999; Knottnerus, Monk, and Jones 1999) has examined the transmission of status distinctions from owners of slave plantations to slave societies embedded within a "typical" nineteenth-century slave plantation in the United States. This study argues that the plantation system had a major impact on the social world of slaves leading to the emergence of social inequalities among them. Based on an examination of various kinds of sources such as slave or owner diaries, autobiographies, travelogues, and historical studies, we find that slave societies were informal groups in which status distinctions were quite common. Oftentimes, the status hierarchy resulted in artisans, house servants, craftspersons, and certain occupations such as valets, foremen, and carriage drivers being accorded the highest esteem, while those who worked in the rice, cotton, or sugar fields received the lowest evaluations.

This investigation of historical sources argues that ubiquitous hierarchical distinctions in status and power in the plantation system (the most obvious distinctions involving ritualized status and authority differences between masters and slaves manifested in different ways in different interaction domains) had a profound effect on slaves as they constructed a social world of their own. Because slaves were continually exposed to and engaged in ritualized practices on the plantation stressing power and prestige differences, this led to a reproduction of ritualized behaviors, relations, and social structure in their informal embedded group. The analysis of how this occurred emphasizes the role of highly ranked RSPs and factors such as salience and repetitiveness for influencing the dominance of these ritual practices.

Another study (Knottnerus and Berry 2002) examines an even older society, ancient Sparta from 600 to 371 BC, and analyzes the social dynamics involved with the training of male youth in this culture.

Drawing on a large number of historical sources, we show how a mode of rearing involving youth embedded within the Spartan city-state and its institutionalized educational program was used to cultivate adult citizen warriors. In studying this very unique society and training system we focus on the various ritualized activities that formed the rearing system. These included the *agoge*—an extremely disciplined, compulsory system of upbringing—military practices, dining clubs, communal living arrangements, and various collective events and practices.

SRT is used to explain how the ritualized practices of Spartan society were reproduced within the social world of youth. The analysis emphasizes how dominant RSPs expressed five core Spartan qualities or symbolic themes: (1) extreme simplicity, frugality, and austerity in behavior and living conditions; (2) social unity, homogeneity, equality, and harmony of group members; (3) hierarchical distinctions and disciplined obedience to authorities and the state; (4) aggressiveness, courageousness, competitiveness, and a militaristic deportment; and (5) secretiveness, deceptiveness, and an opportunistic orientation, particularly in regard to military activities, methods of domination, and propaganda. Ubiquitous ritualized practices emphasizing these qualities profoundly impacted youth and played a central role in the creation and maintenance of the adult Spartan social system, a militaristic social order and way of life that was able to successfully cope with external and internal threats from enemy states and slaves for several centuries.

Moving up in time, Varner and I (2002) have also used SRT to study the emergence of American golf during the late nineteenth and early twentieth centuries. This investigation advances the work of Norbert Elias by providing an analysis of the ritual dynamics involved in the civilizing process as it relates to the game of golf. Using content analysis of texts, such as executive committee meeting minutes of the United States Golf Association, USGA rule books, and golf magazines, we focused on the origins of American golf between 1894 and 1920 and the various attributes of civility that were expressed in this new recreational activity. The study shows that the characteristics of civility found in European noble pastimes were also present to a very high degree in the newly emerging American game of golf. These qualities, which defined the ritualized playing of golf, emphasized a civilized code of conduct and etiquette involving manners, practice, self-restraint, proper decorum, courteousness, compliance, obeying of rules, and honorable behavior. The ritualized enactment of this code of conduct distinguished the upper class (males in particular) from those of a lower status. These practices identified the game of golf as an upper-class activity by socially distancing members of this class from others who were discouraged from engaging in this pastime pursuit.

In short, golf and codes of civility spread from Europe among the nobility to America where these ritualized practices were learned in face-to-face contact in professional golfing associations, clubs, and through golf journals, magazines, and rulebooks. These ritualized practices of civility were in this way ultimately transmitted to all those individuals who constituted upper-class American golfing society.

Finally, Guan and I (1999, 2002) have used SRT to better understand the processes of acculturation and marginalization for Chinese Americans in American society during the twentieth century. Conceiving of an immigrant group as an embedded ethnic group, and communities such as Chinatowns located in certain American cities as embedded ethnic communities, we examine how ritual practices play an important role in the construction of the "unified ethnic character" of the older Chinatown generation and the "biritual ethnic character" of the newer generation of Chinese Americans. We describe and compare these two character types by analyzing the RSPs of Chinese Americans in these different contexts. For instance, in the traditional Chinatowns in American history the members of such communities were highly segregated from American society. They lived for the most part in the self-contained social world of a Chinatown and retained the ritual practices and character of their native Chinese culture.

Different degrees of structural reproduction, however, have occurred among newer generation Chinese Americans, sometimes resulting in a biritual character, which means that actors can easily engage in both American ritualized actions and Chinese ritualized behaviors. Furthermore, various structural, cultural, and global factors have led to differences in structural reproduction and a biritual ethnic character among Chinese Americans. Such a dualistic quality shapes the taken-for-granted lives of people and involves different ritualized activities in a number of social arenas ranging from organizational milieus to more informal social interaction.

Taken as a whole this wide-ranging collection of studies demonstrates how ritual dynamics and reproductive processes can occur in very diverse cultures and time periods and can be studied with different methods. What this work also shows is that ritual dynamics can operate across multiple levels of the social order.

For instance, the laboratory experiment first described involves a task group being influenced by the ritual enactments occurring in another task group. This test provides a demonstration of the basic, or what one might call the generic, ritual process by which structural reproduction occurs. Such a dynamic is at the heart of the process investigated in various studies where embedded groups reproduce the ritualized practices and arrangements present in a wider social environment.

The French school study specifically focuses on how interpersonal relations in embedded youth groups (the first level in the model of social order and ritual dynamics) are influenced by the educational organization (the third intraorganizational level) they are located in. Here the social relationships among youth, and to some extent the wider array of social networks (the second level), are impacted by the ritual dynamics occurring within the social organization of lycées and convent schools, a case of inter-level transmission of ritual practices.

The same applies to the investigation of slave societies embedded in slave plantations. Here, too, the social interaction (first level) and to some degree the social networks (second level) of slaves are influenced by the ritual practices occurring in the social organization of the plantation (the third intraorganizational level). This is another example of the inter-level transmission of ritual enactments.

In the investigation of ancient Sparta we find that the entire community or society (fifth societal level) along with the schooling system (the third intraorganizational level) have a powerful impact on youth, shaping the ritual dynamics occurring in their interpersonal lives (first level involving social relationships). This study provides a vivid demonstration of how inter-level transmission can operate across multiple levels and have profound consequences for people.

Research on the emergence of golf in America widens the analysis by showing how ritualized recreational activities involving ideas about civility can be transmitted at the international/global level (level six) from European societies to America—a type of intra-level transmission—and then influence various organizations associated with golf (third intraorganizational level) and ultimately interpersonal relations and behaviors of golfers (first level). Here we see how ritualized practices that are carried from one global arena to another then undergo inter-level transmission as they spread from the intraorganizational level to social networks among golfers in the more privileged classes and perhaps most importantly the social interactions and relationships among them.

Finally, the study of Chinese Americans involves ritualized behaviors and relationships of people (first level) being influenced by either or both (a) the Chinese ethnic community and associations within it along with more formal organizations involved with this community (second and third levels) and (b) the wider host society of America (fifth societal level). This involves another case of reproduction and inter-level transmission, especially as this relates to the development of a biritual character where the ritual practices of newer generation Chinese Americans are impacted by both their native culture and community and the society in which they live and work.

In saying this, it is important to remember that the ritualized behaviors that occur in people's personal and interpersonal everyday lives (the first level) are addressed in all of these studies. While reproductive processes operate across and within levels among different groups, all of the cases investigated show how these ritual dynamics can affect the behaviors of people in different concrete situations and settings.

Stated somewhat differently, the theory can address these kinds of issues because the formulation conceptualizes social environments and embedded groups in an abstract manner. Consequently, social environments can take many different forms and can be found at different social levels. They can be groups within larger groups, which are themselves within wider social milieus. Reproduction can move through multiple levels from macro to micro levels. At the most micro levels, the focus is on the ritualized behaviors of people that define their personal experiences and relationships in their daily lives.

For example, as golf and civility spread globally from Europe to the United States these ritualized practices affected American golfing clubs, organizations, tournaments, literature, and recreational preferences and in a relatively short period of time were reproduced at the most inter - personal level in the daily lives of those persons playing golf.

Or, in nineteenth-century France, we find that attitudes and practices related to men and women that were prevalent in the society as a whole were reproduced in the workings of boys' and girls' schools. Then the organizational environment of these schools affected the ritualized behaviors and personal relations among youth who were embedded in these settings.

In closing let me return to a point made in the beginning of this chap ter concerning the importance of rituals for human beings. What these investigations show is how potent rituals are. This is the case because, counter to what one might intuitively expect, we find that reproduction occurs in embedded groups even when it does not serve the interests of group members (i.e., there are no incentives or rewards for doing so); when actors in such groups are briefly exposed to ritualized actions in the wider environment; when only some members of the embedded group are exposed to those ritualized activities; when group members verbally express their opposition to the wider social milieu; and when such ritualized practices may result in undesirable consequences for the group.

Findings such as these concerning the power and significance of rituals are derived from the body of work described in this chapter. Grounded in and guided by certain principles within SRT this collection of studies contributes to an expanding, cumulative research agenda. As such, these various studies should be viewed not simply as isolated studies but rather as separate parts of a more encompassing theoretical and research program whose goal is to better understand the role of ritual in human behavior.

Chapter 4

Strategic Ritualization and Power

Nazi Germany, the Orange Order, and Native Americans

J. David Knottnerus, Jean Van Delinder,
and Jennifer Edwards

SRT has now progressed to the point that other lines of research or extensions have begun to develop. All of this research directly or indirectly builds upon the theory's original formulations. Yet, this work directs attention to dimensions of social life and action quite different from the studies examining embedded groups and structural reproduction. Two new emerging lines of research will be discussed in this and the next chapter.

As emphasized since the beginning of the book social rituals play a central role in daily social life. They are a fundamental aspect of human behavior regardless of cultural group membership or historical period. Building on this basic assumption, we find that rituals are often directly connected to social decision-making and power. In other words, rituals are grounded in and shaped by groups who possess the power and orga nizational capacity to control how and why they are used. Indeed, we suggest that rituals and their linkages to social power are widespread, if not universal, occurrences. Such phenomena occur throughout diverse cultures and different time periods. The way this happens, however, can vary widely. Rituals may be used to achieve very different objec - tives, express different symbolic themes and meanings, and take very different forms.

Broadly speaking, the connection between social decision-making, power, and symbolic meanings, or what might be referred to as the cultural dimension of social life, is an important issue that warrants the attention of social scientists. This is a topic that is directly relevant to a number of issues, including the ways cultural factors may facilitate the structuring of social life (Chew and Knottnerus 2002; Knottnerus and Prendergast 1994). Indeed, increasing attention has in recent years been directed to the ways that decision-making and symbolism, culture, or rituals are linked. Researchers in several disciplines have begun to address such concerns, including the ways power, political decision-making, and culture may intersect (see Engelstad 2003; Wolf 1999; Cohen 1969; 1974; 1979; Edelman 1995; Hunt 1984; 1988; Ozouf 1975; Nagel 1997; Hobsbawm and Ranger 1983). Notwithstanding the work of scholars such as these, much more empirical and conceptual research is needed on this topic. This is especially the case in regard to rituals, where we find that the diverse ways in which rituals are influenced by powerful social groups have received limited empirical investigation and analysis within sociology (see Kertzer 1988; Mach 1993; Lane 1981; Petrone 2000).

This chapter, therefore, examines the use of rituals by groups who possess the power and organizational capacity for doing so. To do this, we analyze three cases in very different cultural settings over approximately the last seventy years: the use of rituals by the Nazi party in Germany, by the Orange Order in Northern Ireland, and by Native Americans. In carrying out this cross-cultural investigation attention is directed to three dimensions of the ritual/power relationship: the goals of these groups, the different types of rituals used by them, and the ideas expressed through the rituals supported by these groups.

Strategic Ritualization

The analytical framework of SRT provides the basis for further theory development and research into this issue. One quite relevant line of research focuses on how such practices are sometimes carried out in a deliberate and purposive manner (notwithstanding the fact that ritualization often involves actors engaging in ritualized behaviors in a non-reflective manner). In this regard, it has been suggested that, "Agents can strategically engage in ritualized practices and actively foster the reproduction or transformation of social structures for various purposes including self-aggrandizement" (Knottnerus 1997:275). In other words, "strategic ritualization" refers to people utilizing or manipulating a system of ritualized practices in order to realize certain outcomes.

The idea of strategic ritualization can be wedded to and can build upon the original formulation and research dealing with structural reproduction. This work, discussed in the previous chapter, suggests that the greater the rank of a ritual, as measured by salience, repetitiveness, homologousness, and resources, the greater its impact on the cognitions, behavior, relations, and social arrangements actors are involved in. This is a core analytical argument of SRT concerning ritual dynamics. What this suggests in regard to strategic ritualization is that the higher the rank of RSPs promoted by individuals and groups, the greater the influence those ritualized practices will have on people exposed to these RSPs. Highly ranked RSPs strategically presented to actors should have a significant impact on them, their relationships, and the structure of the groups they are a part of.

Some of the ways individuals and groups may strategically utilize ritualized practices to realize different objectives have been identified in a study of an Italian-American community in the south central United States (Knottnerus and LoConto 2003). Several methodological strategies including interviews, observations, and historical analysis are used to demonstrate how people within this ethnic community have, over the last century, utilized rituals to, among other things, create community bonds, express their ethnic identity, enhance their personal gain, and achieve a meaningful life. Expanding on the original definition of strategic ritualization, a typology is presented of three kinds of strategic ritualization: *ritual legitimators, ritual entrepreneurs,* and *ritual sponsors.*

Ritual legitimators are actors who authorize, validate, or accredit which RSPs are associated with a particular group or collectivity. By employing and indicating support for certain ritualized practices, they attempt to legitimate the practices' relationship to and identification with the social body, whether that be, for example, an organization, a racial or ethnic group, a gender group, a religious group, or a society. Not uncommonly, ritual legitimators occupy a special status or position within a social unit. Examples of this would include a highly esteemed civic, religious, or political actor in society, a leader or "elder" of a group, a person considered to possess a special expertise in an organization, or the successful member of a well-known and respected family within a community. The judgments, opinions, and evaluations of such persons often carry more weight than those of others in the group. And not surprisingly such persons tend to be trusted by others.

These individuals are perceived by others as possessing a special knowledge, understanding, or insight that gives them the right to decide the value and meaning of RSPs associated with a collectivity. Various kinds of knowledge are possible. For instance, it could involve well-known religious or political leaders who are perceived as possessing a vision or special understanding of events, customs, and conditions. It might involve older members of a community who are viewed as having a privileged understanding and insight into the past, such as an elder in a Native American tribe or a long-time member of an ethnic community in a town. If others wish to acquire this special knowledge, ritual legitimators are the individuals people must communicate with. By validating or legitimating the RSPs associated with a social body, these persons help to identify the particular qualities that distinguish a group and the moral worth of that collectivity.

Ritual entrepreneurs are actors who employ RSPs associated with a particular group or collectivity in the carrying out of economic activities. In order to facilitate and promote some kind of economic or market venture actors use ritualized activities that are or have been associated with a certain group. Ritual entrepreneurs may be engaged in many types of economic activities and employ RSPs associated with racial, ethnic, religious, gender, labor, political, or other sorts of groups. This type of strategic ritualization could, for instance, involve individuals who run smaller-sized business operations. Examples could include small businesses such as bakeries, delis, clothing stores, restaurants, arts and crafts shops, and food markets, selling items associated with the ritualized eating, clothing, or artistic practices of a group. The opening in a community of an Italian or Thai restaurant, an Asian grocery store, or the development of a specialty shop selling artistic items, porcelain, or jewelry associated with Native Americans illustrates this type of activity. Larger-scale economic operations such as corporations also use ritualized practices associated with a group to market and advertise different products. The scope of the economic operations and ritualized activities may extend to an entire society or throughout the world. Examples of this practice would include the global merchandising of certain forms of music associated with a particular country or an ethnic group (e.g., reggae with Jamaica or blues or rap with originally American singers).

It is important to note that while the goal of this type of strategic ritualization is economic profit, it is possible that other objectives exist that coincide with these economic motives. A business, for example, that engages in this type of behavior may at the same time have a strong identification with the ethnic or racial group it associates itself with. It may be quite committed to the group and its ritualized practices. The ritual entrepreneur may be viewed as a carrier of the torch for an ethnic group's rituals, history, and collective identity. So, their motivations may not only be economic in nature, but might also involve the desire to contribute to the social and economic well-being of the group, while also seeking to represent the group in a favorable manner to others, such as the visitors or tourists who come to their community. An example might be a successful and well-known Italian restaurant in an Italian American community that is committed to the preservation of that ethnic community and how it is perceived by others in the wider society. In this case, ritual entrepreneurs may provide some degree of legitimacy for the RSPs related to that group.

Ritual sponsors are individuals who develop and promote events comprising RSPs that are representative of or associated with a particular collectivity. By sponsoring such events, certain RSPs associated with that group are made more prominent or conspicuous and more accessible to members of the group and possibly to non-group members such as the public at large. The reasons ritual sponsors promote and facilitate certain events vary. Motives for engaging in such behavior could range from political, civic, or economic objectives to the need for public attention and recognition or the desire to develop and project a particular social identity and sense of uniqueness or distinctiveness for the group. Aside from the singular or multiple outcomes sought by ritual sponsors, the sponsoring of events by actors is usually considered a legitimate type of activity, occurs within existing institutional arrangements, and utilizes what group and non-group members would consider to be reasonable and appropriate procedures and tactics.

Various kinds of events can be promoted by ritual sponsors, such as festivals, dances and musical activities, pageants, commemorations, celebrations, political rallies and special meetings, or religious events. A common example would be a festival in an ethnic community. Here, numerous ritualized practices associated with the ethnic group and its nation of origin may be engaged in, displayed, and celebrated. Ritualized activities that are found in ethnic festivals could include the preparation and eating of ethnic foods; the playing of ethnic music; certain styles of dancing; wearing special clothes; symbolic markers for the group such as the display of banners, pictures, flags, or logos; playing traditional sports or games; religious services and other practices associated with the ethnic group; and the display or sale of certain types of crafts, artwork, or other items identified with the group.

Both group members and non-group members may act as ritual sponsors. Sometimes such events, which can be quite large, may require the cooperation and assistance of non-group members from the wider community. This assistance may be of a civic, legal, political, or even economic nature. Perhaps not surprisingly when such assistance occurs this may lead to changes in the RSPs that make up the event. It is possible that ritualized practices not associated with the group will be introduced into the event. At the same time practices associated with the original group will probably continue, although they may be transformed to varying degrees. What these changes involve depends on various circumstances. For instance, some traditional customs associated with the original group sponsoring the event may weaken or disappear, while other practices may become more exaggerated and prominent. These changes could involve ritualized activities and images that group members wish to stress and communicate to others and/or that are in line with stereotypical beliefs about the group in the wider society. It is quite likely that this has occurred over time in some ethnic festivals in the United States where practices such as eating distinctive foods, playing certain kinds of music, and selling specific kinds of beverages have increasingly dominated these events.

Having defined these three types of strategic ritualization—ritual legitimators, ritual entrepreneurs, and ritual sponsors—I would point out that while they are conceptually distinct in reality it is quite possible that people could exhibit features of several of these types. And they could do so at the same time. Stated somewhat differently, empirical cases can include the characteristics of more than just one type of strategic ritualization, although normally we would expect that the features of one of these three types would be most evident in any particular situation. Having said that, these concepts enable us to recognize and understand some of the ways people strategically engage in rituals to achieve various ends.

While these three types of strategic ritualization are often found in societies these analytical concepts still do not exhaust this topic. A fourth type, not previously discussed, is also quite common and particularly relevant to the present chapter. The fourth way people may strategically employ rituals involves ritual enforcers.

Ritual Enforcement

Ritual enforcers are actors who use social power and/or a position of domination to determine the RSPs that are associated with a particular social unit. In other words, they directly influence or control the kinds of ritualized practices that occur within a group and that people are, therefore, engaged in or exposed to. Such power may be exercised in an overtly authoritarian and coercive fashion or may be used in a more subtle and nuanced manner and be less visible. Be that as it may, power differences grounded in, for instance, social organizations, communities, or entire societies are what allow ritual enforcers to strategically determine the occurrence of ritualized practices.

The use of social power to control ritualization processes may be influenced by the desire to bring about various outcomes. Such goals could range from economic gain to the desire for prestige, social honor, social identity, a sense of moral worth, and acclaim. But underlying and permeating the attempt to achieve any and all of these ends is the fundamental motivation to control people's ritualized behaviors and cognitions. Indeed, the primary motive for individuals attempting to influence the ritual enactments of others is oftentimes the desire for power itself, especially political power.

Ritual enforcers may take various forms and operate at different social levels. At the micro level, instances of ritual enforcement could include efforts by individuals to strategically control the ritualized practices actors engage in while interacting with one another, for example, controlling rituals of deference or ritualized behaviors denoting authority in face-to-face interaction. At a more macro level, one might find certain individuals in a community or organization manipulating and controlling various kinds of collective ritualized events. Such enforcement could include efforts aimed at preventing or ensuring that certain kinds of political meetings, civic events, religious services, social gatherings, festivities, ceremonies dedicated to those who have died, and celebrations memorializing persons and events occur in a social setting. At an even more macro level, similar processes and forms of influence could operate throughout an entire society. Examples would include the control and manipulation of large-scale ritualized practices such as mass rallies, special holidays, or commemorative events in authoritarian, totalitarian, or single party systems such as the Soviet Union or fascist Italy during World War II.

In addition to the notion of ritual enforcer, we also introduce another concept—program of ritualized symbolic practices—that is directly relevant to our concern with how groups may tactically utilize rituals to achieve various goals and express important ideas, that is, schematic meanings or symbolic themes. The concept of *program of ritualized symbolic practices* is defined as a collection of RSPs strategically used by a group or individuals such as group leaders to achieve certain objective(s). This concept directs attention to the fact that groups or organizational units may have the differential capacity to employ an assortment of ritualized practices to achieve one or more end(s), including the maintenance or enhancement of the group's power relative to other social entities.

The ritual practices that compose a program may vary in terms of their number, nature or type, focus, frequency engaged in, and other significant features. The practices within a ritual program may also complement one another to differing degrees in terms of the meanings of the symbolic themes expressed through them. Overall, ritual programs can vary in their degree of complexity. Programs could, for instance, comprise numerous ritualized practices, each of which is significant and complements the others. On the other hand, programs may be simpler in nature, defined by only one or two dominant ritualized symbolic activities and possibly other practices of lesser import. This concept directs our attention to the great variations that may occur in the ritual programs supported by groups and the enormous differences that exist in the degree to which groups plan and draw on resources to construct ritual programs.

In essence, the present chapter represents the first effort to identify and describe some of the varied ways in which quite diverse cultural groups, including key leaders, have in different time periods strategically enforced ritual programs of varying degrees of complexity to achieve their ends. Drawing upon the power base and organizational capacity of these groups, rituals as we shall see have been employed to achieve various outcomes. In conducting this investigation, attention is directed as previously mentioned to three key aspects of the ritual/power relationship: the objectives of these groups, the different kinds of rituals used by them, and the ideas or themes expressed through the rituals implemented by these collectivities. The first of the three cases we examine provides the most comprehensive instance of a program of RSPs directed toward an entire society.

Nazi Germany

The dynamics surrounding social ritual and power are clearly evident in one of the, if not the, most studied events in the twentieth century: the rise of Nazism spearheaded by its leader, Adolph Hitler (for discussions of the use of rituals in another powerful political system, the Soviet Union, see Lane 1981; Petrone 2000). As Kertzer (1988:163) states: "Although all political movements become known through their rituals and symbols, the identification of politics with ritual is perhaps nowhere more graphic than in the case of Nazism." At the same time, despite the many excellent historical treatments of Nazism "there have been surprisingly few attempts by social scientists to examine the political celebrations developed by National Socialism" (Taylor 1985:175). As Taylor notes, only H. T. Burden's (1967) book on the Nuremberg rallies addresses this general issue, providing an extensive description of these rallies but not examining any other types of Nazi political celebrations and providing a very limited analysis of the rallies themselves.

Other than that and Taylor's own work (1981; 1985, especially chapter 7) one must turn to two German authors (Vondung 1971; Gamm 1962) for any kind of analysis of Nazi ritual and its connections to the broader dimensions of Nazi society. Gamm (1962) focuses on the cult of the leader and the religious symbolic quality of the Nazi movement. Some attention is given to the ritualistic planning of free time, for example, summer camps for Hitler youth, raising of the flag, and yuletide celebrations. Vondung's (1971) analysis places greatest attention on what is perceived to be magical elements in the political religious cult of National Socialism. In this analysis some discussion focuses on nationalistic celebrations during the course of the year and how celebrations occurred in special places and might utilize sacral language and music.

Notwithstanding the limited attention formally directed to this topic, various sources and insightful discussions of National Socialism make it possible to identify in a much more focused and comprehensive manner the goals of Nazism, the symbolic themes expressed by its rituals, and the various kinds of ritualized practices this group employed (for a small sampling of works dealing with Nazism and its leaders, see Zeman 1964; Fest 1973; Herzstein 1978; Baird 1990; Bullock 1992).

As a form of fascism, the National Socialist German Workers' Party (NSDAP), like other European fascist organizations of this period, thought of itself not as a political party in the usual sense of the word but rather as a political and social movement. It (and other fascist groups) has been described as "an antimodernist protest movement" against representative democracy, modern society and pluralism, the capitalist economic system, and the ideology of economic and political liberalism (Beck 1997:623). It self-consciously identified itself as a revolutionary movement attempting to achieve the goal of fundamentally altering the social, political, and economic structure of German society. Thus, the Nazi movement sought to "re-educate" the German people for a new society grounded on a new *Weltanschauung* or "revolutionary" belief system. By exposing the people (or *Volk*) to a new value system focused on the purity of the Aryan race, the National Socialist State sought to restore a true consciousness to the members of society, thereby contributing to their spiritual and moral well-being.

Stated somewhat differently, Mosse (1964; 1975), who has arguably provided us with the most sophisticated analysis of *volkisch* thought and mass national movements such as the Third Reich, describes the Nazi political movement as a secular religious movement. It was a "secular religion" of nationalism defined by an array of rituals, symbols, festivals, and other collective activities (Mosse 1975:214). Welch (1993) partially summarizes this thesis concerning the essential components of *volkisch* thought. In pointing out that Nazism, rather than being an aberration in the history of a society, was based on elements of intellectual thought that go back at least a century, Welch (1993:52) emphasizes that the major themes that recur in the Nazi belief system "reflect the roots and antecedents of *volkisch* thought: (1) appeal to national unity based upon the principle of 'The community before the individual' (*Volksgemeinschaft*); (2) the need for racial purity; (3) a hatred of enemies which increasingly centered on Jews and Bolsheviks; and (4) charismatic leadership (*Führerprinzip*)."

Focused primarily on the *Volksgemeinschaft,* "an organic national and racial community" (Diephouse 1983:56), National Socialism aspired to a "spiritual revolution" (Mosse 1964:292) that would include all Aryans. By rediscovering their Aryan roots and defending against "racial pollution" the Volk would achieve "racial purification" and a "spiritual strength and purity" (Wolf 1999:238, 239, 236). And, consistent with the belief that life fundamentally involves an unremitting struggle, the attainment of a true *Volkhood* was to be brought about through the effort and guidance of an extremely committed and ruthless political elite (for further discussion of the ideological and intellectual arguments that make up this belief system, see Mosse 1964; 1975; Diephouse 1983; Wolf 1999, especially 235–42, 266–73; Zentner 1997; Jäckel 1972). To a large degree then, one could say that Nazism and its political leadership were marked by a drive for "moral" and "revolutionary purity," which sought to overcome and/or eliminate all other groups defined as morally inferior, impure, and polluting (Moore 2000).

The RSPs engaged in and promoted by the Nazi party articulated symbolic themes consistent with these ideas. Ritualized activities celebrated the superiority of the community, or *Volk*. They expressed support and adulation for the leaders of the movement, the Nazi party, and most importantly the Führer who was often accorded absolute power and was oftentimes viewed in a messianic manner. They communicated themes that emphasized the distinctive and superior worth of their group while negating, degrading, and supporting the destruction of others, especially those considered most "impure." And these ritual practices expressed a number of other related symbolic themes that were logically connected to, consistent with, and/or supportive of these basic or general concepts.

Such themes included the need for social conformity to the national *Volk*-community, its essence, and the attributes and behaviors through which it was manifested; a warrior/military orientation involving characteristics such as strength, discipline, control, obedience, perseverance, and training that were best suited to a life conceptualized as struggle, to becoming a master race, and to defeating one's enemies; unity of the people and its leader; a religious/sacred quality to the movement and its leader; harmony, cohesion, and unity of the *Volk*-community, which was sometimes conceptualized in utopian terms; and sacrifice for the community, the movement, and the Führer.

These symbolic meanings were expressed through a wide variety of quite different ritualized practices. Indeed, the extent and complexity of these various practices were so pronounced that it is not possible to discuss all of them here. We will, therefore, briefly describe a number of the most common RSPs in order to convey a sense of the scope of the activities that were sponsored and enforced by this nationalistic movement.

Probably the most well known and dramatic ritualized event was the Reich Party Congresses (Burden 1967). First staged in Munich in 1927, from 1928 to 1938 these large mass rallies were held in Nuremberg at the beginning of September and lasted for a week. The grounds where they took place were constructed in a grandiose style and held huge numbers of participants. Formations of several hundred thousand persons could march on the field. The field was bounded by ramparts containing stone turrets with banners on the top along with a main platform and a very large portico. A stadium, parade ground, and large boulevard were also part of the field. As various commentators have noted (e.g., Taylor 1985; Kertzer 1988), these were monumental architectural structures for celebrating National Socialism. A clearly defined set of ritualized practices occurred at all rallies.

Rallies began with memorial services for those killed in World War I—associating Nazism with their sacrifice—and incredibly large columns of uniformed and flag-bearing members of the SA, SS, Hitler Youth, and Nationalist Socialist Motor Corps, who passed by Hitler (Wehrmacht units were also included). During the week ceremonies dedicated to the dead were held, sports demonstrations were staged, symbolic objects such as banners were sanctified, and oaths of loyalty were expressed. These activities led up to the major speech by Hitler at the end before huge numbers of people assembled on the grounds. Countless red banners and spotlights streaming their lights in an orchestrated manner above the assembly accompanied these events. At the center of and above the massive columns Hitler stood alone on a stage where he delivered his speeches through a carefully prepared speaker system. These events created a magnified sense of community, unity, obedience, and military bearing and devotion among participants, while highlighting other key symbolic themes involving the exceptional nature and absolute power of their leader. The entire proceedings assumed an extraordinary if not sanctified quality. Finally, in his speeches, Hitler consistently articulated and dramatized various themes ranging from the superiority of the Aryan people and the Nazi movement and the need for conformity to and sacrifice for the cause, to the dangers posed by enemy groups and the need to overcome these adversaries. In total it was a massive and quite moving ritualized collective event shaped and defined by an array of carefully designed practices.

In addition to these massive, dramatic rallies, the Nazi party sponsored many other types of ritual practices in German society. Actually, large rallies were just one of many celebrations in the official National Socialist calendar. Various party units such as the Propaganda Ministry and the Reich Propaganda Directorate organized these holidays and memorial days. For instance, the following celebrations occurred throughout the calendar year (Zentner and Bedurftig 1997:132–33): January 30, Day of the Seizure of Power; February 24, Party holiday in honor of the proclamation of the Party Program of the NSDAP in 1920; March, (1) Remembrance Sunday, the Heroes' Memorial Day, held on March 16 to commemorate the introduction of universal compulsory military service in 1935, (2) Commitment of Youth on the last Sunday of the month involving the acceptance of fourteen-year-old children into the Hitler Youth and League of German Girls; April 20, Führer's Birthday; May 1, May Holiday; different days in May, Mother's Day; June 21, Summer Solstice; September, Reich Party Congress in Nuremberg; October, Harvest Thanks Day; November 9, Memorial Day for the sixteen "blood witnesses of the Movement" in the Hitler Putsch of 1923, which was the highest National Socialist Holiday; and December 21, Winter Solstice, the [unsuccessful] attempt to replace Christmas with a feast.

Some celebrations, especially in the early years of the Nazi movement, built upon and transformed preexisting rituals. For instance, May Day (centered on working-class allegiance to the socialist movement) was changed to the Festival of National Brotherhood honoring national solidarity within the Nazi social system. And March 16, Remembrance Day (honoring World War I dead) was transformed into Heroes' Remembrance Day, which emphasized the importance of the military. So too, traditional folk festivals such as the harvest festival were transformed into official holidays. While hundreds of thousands of farmers in very large festivals listened to Hitler and others speak in Westphalia, smaller versions of this ritual occurred throughout the country.

Another type of collective event involved ritualized political rallies. Political rallies were a central vehicle for the dissemination of Nazi ideology and propaganda. Although Hitler had stressed the significance of the spoken word and was by far the most moving speaker, "great effort was made to train large numbers of speakers for the Nazi movement, and special courses were set up to teach speakers to address Party rallies" (Bessel 1980:25–26). Not surprisingly, from 1930 onward the NSDAP surpassed other parties in holding public rallies.

In addition to the ritualized presentation of annual events and holidays, political rallies, and mass rallies, the Nazi party sought to substitute the National Socialist "morning celebration" for the Sunday services of churches. In these gatherings "Decor and programme were purposely restrained to evoke an atmosphere of pathos and solemnity. Against a background of selected music by classical or contemporary 'accepted' German composers a Hitler Youth choir declaimed an ode to the glory of the 'Thousand Year Reich,' extracts were read from the speeches and writings of the Führer and other Nazi luminaries and the ranking party leader gave a brief but impassioned address" (Unger 1974:175; for an eyewitness account of how political festivals were turned into quasi-religious festivals, see Herman 1943).

Unger (1974:176) also points out that "a direct onslaught on the last preserves of privacy was represented by the so-called 'life celebration.'" Building on traditional customs, rituals centered on the *Volk*-community, party, and Führer were developed dealing with birth, marriage, and death. To the extent that normal citizens engaged in these practices, the party was able to promote RSPs that had relevance for the most intimate occurrences in the lives of individuals and families. In 1944 the Nazis even initiated a "National Socialist family evening" in which the local party leader was to arrange "family evenings that were 'lively [and] colourful,' preceded by music and songs and followed by a 'brief political speech'" (Unger 1974:178).

In this same spirit the Nazis attempted to integrate ritualized recreational and cultural practices into the political movement. Thus, the Labour Front's "Strength through Joy" movement promoted holiday camps and excursion boats for citizens. Furthermore, "all the various forms of communal recreation from folk dancing, amateur drama groups and sport competitions, to collective theatre, cinema and concert outgoings were fostered and controlled by the local leader of the party in the drive to draw the people away from 'non-committing private amusement' and into 'the festive community'" (Unger 1974:171). The use of a wide array of collective RSPs such as these were clearly aimed at monopolizing as much as possible the personal lives of people and involving them in ritual events associated with the Nazi party and its ideas.

Although different in form, the goals of integrating people into and conforming to and sacrificing for the community were also realized through ritualized programs that provided material aid to fellow citizens. These included the "Winter Help" program, which collected money and material goods for impoverished families, and the "One-Pot" meal, which encouraged families once a month during the winter to donate part of their Sunday lunch to others. As Welch (1993:60) notes, these public "rituals ... were intended to represent a vivid expression of the newly created 'national community' and proof of loyalty to the regime."

These symbolic themes were also expressed and reinforced through a ubiquitous array of additional ritual activities. Such practices ranged from continuous commemorations for those killed in war to the singing of the national anthem when Nazi symbols such as the nation's red flag with swastika were shown. In essence, Nazi rituals enveloped daily life (Kertzer 1988:167). Members of the party had to continuously wear the Nazi insignia and wear party uniforms in public gatherings. And social interaction was normally characterized by citizens greeting one another with a Nazi salute and reciprocating with "Heil Hitler," a practice that if ignored could result in severe sanctions.

In sum, a number of clearly defined symbolic themes were forcefully communicated through an extensive program of ritualized activities strategically cultivated and disseminated by the Nazi party apparatus. These practices ranged from quite prominent (i.e., salient), periodic collective events to other types of rituals that, while less dramatic in nature, were engaged in quite often if not almost constantly (i.e., high degrees of repetitiveness). So too, these activities varied in terms of their being more formally organized and larger in scale or smaller and more personal in nature. In other words they occurred at many different levels of the social order ranging from the micro level of social interaction to more macro levels of organizations and interorganizational relations and ultimately society as a whole. Taken together they compose a strikingly comprehensive collection of dominant or highly ranked RSPs that, notwithstanding certain problems surrounding their implementation (e.g., competition among some party leaders and agencies for the control of rituals and some officials' incompetence in facilitating ritual practices), undoubtedly contributed to the enthusiastic support among some and, at the very least, tolerance among many for Nazism and its leaders.

The Orange Order

For more than two centuries, the Orange Order has been a major force in Northern Ireland. The dominance of this group is expressed most clearly in a very special and quite prominent ritualized practice: the Twelfth of July Parades. Through the annual practice of parading, the belief system and social prominence of the Orange Order are symbolized and reproduced by its members.

The Orange Order, a Protestant and patriotic organization, was established in honor of Prince William of Orange due to his victory over King James II at the Battle of the Boyne on July 12, 1690 (actually, the Battle of the Boyne occurred on July 1, 1690, but the Orange Order commemorates the battle on July 12 due to the final victory of Protestants over Catholics during the Battle of Aughrim on July 12, 1691). The Orange Order was not, however, officially established until 1795 after the Battle of the Diamond. Following the battle, Protestants gathered together and declared their brotherhood in loyalty to the crown, the country, and reformed religion (www.grandorangelodge.co.uk). According to members of the group, the primary purpose of the development of the Orange Order was to preserve religious freedom for Protestants. Modeled after the Masonic Lodges, the Orange Order emerged during a time of sectarian friction as a religious organization (Cecil 1993).

Today, the Orange Order remains active in Northern Irish politics, committed to Protestantism, British rule, and perpetuating Unionist power in Northern Ireland. The goal of the Orange Order is to articulate and assert its Protestant beliefs, presence, and loyalty to Britain. Central to the strength of the Orange Order is the hierarchical organizational structure of this group. Members in each level of the organizational hierarchy communicate and work together, maintaining the strength of the order.

More precisely, leaders of private lodges must communicate and coordinate activities with leaders of the district lodges. Leaders of the district lodges then communicate with officers of county lodges, who in turn communicate with the grand lodge. Information is then relayed back to private lodges throughout the regions of Northern Ireland. Thus, while each level of the hierarchy maintains a certain level of autonomy, all units generally work together, as they engage in various activities, especially certain ritualized practices (for useful descriptions of the Orange Order, see Farrell 1980; Jarman 1996; Haddick-Flynn 1999; O'Malley 1995; Darby 1997). Moreover, the Orange Order carries out RSPs in a deliberate and purposive manner in order to produce certain results (Edwards and Knottnerus 2007; see also Edwards and Knottnerus 2010). As Edwards and I state: "The Orange Order is organized in such a way that private lodges in particular as well as district and county lodges have the opportunity and capacity to develop and strategically carry out ritualized practices in many communities throughout Northern Ireland" (2007:190).

Each year, members of the Orange Order participate in numerous ritualized activities, such as church services and high tea, that celebrate the Twelfth of July. The most important ritualized practices, however, are the Orange Parades. Approximately three thousand parades occur each year. According to Bryan (2000), members of the Orange Order participate in several types of parades. Specifically, Orangemen and Orangewomen march between April and August in parades such as arch openings, church parades, mini-twelfth parades, commemorative parades, and the Twelfth of July Parade. This description and categorization of these rituals, we would note, is based not only on secondary analysis of sociological and historical evidence but, most importantly, on direct observation of Orange Order activities and interviews with members of the Order in 2001.

Arch opening parades represent a celebration of the lifting of an arch, which is decorated with Protestant symbols. Additionally, church parades, mini-twelfth parades, and commemorative parades represent important ritualized practices for members of the Orange Order. Church parades occur before and after a church service. Here, members of the Orange Order march from a gathering place to the church for religious services (normally a short distance). Following the church service, participants march back to the original meeting location. Mini-twelfth parades occur prior to the Twelfth of July in celebration of and preparation for the large Twelfth of July Parade. Commemorative parades serve as a ritual for remembering historical events in Britain such as World War I (the 36th Ulster Division). Finally, the Twelfth of July Parade, commemorating the victory at the Battle of the Boyne, represents the most significant RSP of the Orange Order (Bryan 2000).

Although there are different types of parades, all parades share similar characteristics, including the type of music played (with the exception of church parades, which, according to members of the Orange Order, usually include hymns), clothing worn by members of the Orange Order, the display of banners (with the exception of church parades, which normally do not include banners), and parading style. Generally, Orange Parades consist of a march down a traditional parade route, extending over most of the Protestant community; although some differences do exist, involving, for instance, the Orange Order marching a short distance to and from a church service. Top Orange Order officers march in the front of the parade with different bands playing patriotic and religious music followed by members of various Orange lodges. Clothing worn by members of the Orange Order includes a suit, tie, and bowler hat (females generally wear a skirt and jacket). All Orange Order participants wear a sash, perhaps the most significant symbolic aspect of the parades.

Possible colors of the sash are orange, blue, purple, and red, with each individual lodge choosing the color of its sash.

During the parades, some members of the Orange Order carry banners (again, banners are not carried during church parades, as these parades are short and it is a time of worship). The banners represent a very important aspect of the parades, as various religious and patriotic scenes are depicted. Further, some banners contain images of deceased members of the Orange Order. In addition to banners, British flags are often carried, symbolizing loyalty to Britain.

Several symbolic themes are expressed in the ritualized practice of the Orange Parades. Analysis based on observation, interviews, historical sources, and examination of other forms of evidence such as written materials disseminated by the Orange Order and published ethnographic accounts of the group's activities reveals that the most prominent or salient themes communicated by the parades center on the ideas of dominance, British loyalty, opposition to Catholics, and moral superiority. These themes are most clearly evident in the Twelfth of July Parades in Belfast.

The parades express quite distinct themes of dominance/control and loyalty to Great Britain. They provide an occasion for celebrating the memory of a three-hundred-year conflict between Protestants and Catholics in which Protestants have achieved a privileged political, social, and economic position relative to Catholics (Jarman 1997). The importance of the Orange Order is, in large part, due to its loyalty to the British Empire. In essence, Unionists (Protestants) convinced the British government that they were a legitimate part of the British state and not just Irish. Furthermore, they redefined Ulster, the center of Unionism, as being part of British culture rather than Irish culture. Blending elements of both high culture and folk tradition, the Twelfth of July Parade is a symbolic ritual that articulates and celebrates themes of loyalty to Protestantism and Britain (Jarman 1997). And although the social climate in Northern Ireland has changed since the data were collected, these changes have not impacted the organizational structure or processes of strategic ritualization. Hence, the Orange Order has remained politically active, playing an important role in promoting Unionism in Northern Ireland (www.cain.ulst.ac.uk/issues/parade/organis.htm).

Stated somewhat differently, the Orange Parades have become a public ritual in a complex society, thereby providing a medium for symbolizing Protestant power. Gestures of dominance are expressed during the parades as the Orange Order marches through predominantly Catholic neighborhoods (Kenney 1991). For instance, Protestants normally march in Cookstown, a town with a Catholic population that has for years protested the marches. In this regard, the Orange Order believes its power is manifested through its being free to march in the way in which it chooses. Protestants have repeatedly declared that they will not be told when and where they can march (Kenney 1991). By the very fact that the parades occur in both Protestant and Catholic areas, they symbolize Protestant domination of Catholics and are celebrated as a victory over them.

For this reason it is evident that both ritual and politics are integral components of the annual event. Indeed, it may be argued that both politics and ritual contribute to the conflict and violence in Northern Ireland. In this vein, Rappaport (1979) contends that expressions of political power on the local level (e.g., Orange Parades in Belfast) influ ence outcomes on the national-international level, thus explaining the role of parades in fueling political violence and determining the course of political events.

At the same time this ritualized practice serves to bring people of the same faith closer together, thus dramatizing the differences between Protestants and Catholics and intensifying the social distances between the two groups (Jarman 1997). Not surprisingly, social relations in Northern Ireland are often referred to as the Protestant/Catholic conflict, reflecting the intense opposition and high degrees of separation between Catholics and Protestants. At the heart of this condition lies the ritualized behavior of parading that symbolizes oppositional themes and collective loyalty to one's group whose members are expected to unite together against its enemy.

Finally, moral superiority represents another distinct theme expressed in the Orange Parades. In these annual ritual events the moral worth or value of the Orange Order and Protestantism is constantly stressed. Here, moral worth refers to a collection of beliefs involving a differential evaluation about the value or desirability of a group. One finds that ideas about the moral worth of the group are regularly voiced in these ritual events (Jarman 1997). For instance, the presumed moral worth of the Orange Order is often demonstrated in their expressed belief that they are religiously superior to other groups. A reverend declaring that God is on the side of Protestants in a speech at a Protestant assembly provides a simple example of such behavior (Kenney 1991). Other examples would include speeches at rallies referring to Catholics as "Republican scum"

(Kenney 1991), a characterization that amplifies the notion that such group members are morally inferior.

By supporting and repeatedly engaging in this program or collection of highly dominant RSPs, the most important of which involves the Orange Parades, members of this highly organized political-religious group attempt to bring about various outcomes. Some of these are clearly connected to the symbolic themes communicated through these ritualized activities. Such ritual enactments contribute to a heightened sense of moral worth among members of this group and many other Protestants not officially belonging to the Orange Order. The degree to which such ritualized practices are engaged in, their dominance, and their very nature (e.g., occurring in Catholic areas), as pointed out, also provide distinct messages concerning the power of this group in Irish society.

Moreover, this ritual program heightens the status of this group. By status we refer to the Orange Order's position in the social prestige system of Northern Ireland. Participation in these various rituals has directly contributed to the high prestige of the Orange Order and Protestants in general in this society (at least from their own perspective). By continually engaging in dominant or highly ranked ritual activities that demonstrate political dominance, an on-going association with the British Empire, and perhaps most importantly moral superiority, such practices have repeatedly served to reaffirm Protestants' belief in their elevated status.

Lastly, the ritualized practice of the Orange Parades contributes to the strengthening of the collective identity of the Orange Order. By collective identity we mean the similarity of beliefs, ideas, and behaviors among members of a group resulting in a shared collective self-representation and a sense of community, especially as they relate to past and current group events, members, and their common attributes (our definition).

It is quite apparent based on interviews, field observations, and an analysis of secondary sources that the various components of ritualized parading, and to a lesser degree other ritualized practices such as rallies and meetings, involve shared activities and ideas that result in a collective representation or identity concerning the distinctive qualities and history of this group. The collective identity of the Orange Order is, among other things, enhanced by their ritualized clothing styles, music, murals, and banners used to display messages about the group and their Protestant heritage. For instance, the banners carried during the parades highlight beliefs, historical developments, and individuals associated with Protestantism. As a result, a collective self-representation emerges, grounded in a sense of community defined by history and a sense of independence (Jarman 1997).

In sum, the Orange Order, a major political-religious group in Northern Ireland, expresses its beliefs through the Twelfth of July Parades. Ritualized practices such as these that occur several thousand times each year have contributed to its being a dominant force in this society and a key player in what has seemed to be an irreconcilable conflict between Protestants and Catholics (Todd 1994). As a highly organized, hierarchically structured group with a strong and effective leadership, ample resources (e.g., economic assets, large numbers of members, etc.), a long and proud heritage, and the capacity to mobilize its membership, the Orange Order has successfully supported a program of RSPs aimed at achieving various objectives including the advancement of its political power and the collective identity and solidarity of group members.

Although certainly quite different from the Nazi party in its substance and goals, definitely not as extensive in scope as the ritual program developed in the Nazi system, and not involving a totalitarian state regime, the Orange Order has still managed to create a very effective, highly organized ritual program that has become an integral, institutionalized part of this group's endeavors for an extended period of time. Centered to a large degree on one type of ritualized activity—parading—the Orange Order has effectively utilized its organizational and political capacities to carry out this ritual practice. In acting as a ritual enforcer it has been able not only to ensure the enactment of this practice within its own organization and the religious population it represents but also to impose it on a rival religious group while preventing the latter from engaging in a similar ritual activity in Protestant territories.

Native American Powwows

We now turn to a very different type of activity: contemporary Indian powwows in the United States. Powwows are a ritualized practice that provides indigenous peoples with a way to maintain cultural autonomy within a society that has continuously tried to subdue them politically, economically, religiously, and socially (Lerch and Bullers 1996; Mathews 1999; Young 1981). Here we examine powwows in present-day Oklahoma (for video recordings of such rituals see the videography section at the end of the list of references). This discussion extends SRT and the concept of RSP to focus on a domain of power and resistance sometimes overlooked in the study of cultural systems: the body. In this section, we discuss how American Indians use their bodies—particularly through dancing, singing, and drumming—to renew their culture, cultivate feelings of unity, and resist acculturation.

The contemporary powwow is a form of expression that is "culturally innovative" and is "not only intertribal, but also a mixture of sacred and social dances, [which] may contain spiritual, cultural, political, and economic dimensions" (Champagne 1999:127). Indeed, Mattern (1999:129; see also Lerch and Bullers 1996) argues that contemporary powwows are a crucial "public arena" in which contemporary Indians negotiate tribal and intertribal unity as well as individual identity. For these reasons powwows offer another important way to study how groups engage in ritual programs.

The term "powwow" originates from an Eastern Algonquian word that came to be used in English to refer to "any Indian gathering" or as a verb meaning "to confer in council" (Heth 1992:105). For Native peoples today, it means a "secular event featuring group singing and social dancing by men, women and children" (Heth 1992:105). As such, powwows play a key role in the sociopolitical, religious, and cultural revival of American Indians (Call 1991; Gelo 1986; Sanchez 1995). They are "significant cultural events" that are "especially important in their assertion of cultural continuity" (Iverson 1999:32). Powwows also offer a visual narrative of the persecution and survival of Native American resistance to the dominant Anglo culture (Axtmann 1999). As we will discuss, particularly through the ritualized practice of dance, body movement is used to transform everyday activities such as telling stories, drumming, and singing into invisible tactics of empowerment and group affirmation (de Certeau 1988; Lassiter 1999).

As a program of ritualized practices, powwows offer a powerful medium for native peoples to cultivate a distinct ethnic identity and solidarity in the face of bigotry. In this regard, Kracht credits powwow dancing as the crucible through which Indian ethnicity is continually cultivated and renewed. In this dynamic and creative process of Indian identity and pride:

> powwows have become an integral part of intertribal social life ... [t]he common celebration of Native American traditions in today's powwows has bridged history with future survival, and the spirit of the dancing tradition is sacred to many. [It is] ... a continuum in Indian life.... We

> live this spirit on a daily basis. It is why we have survived for so long. At one time we were a forgotten people, but I think we are getting stronger. From the powwow we can gain strength as Indian people, individually and collectively, to go on into the 21st century. (1994:322)

Although resistance to assimilation expressing and affirming qualities and characteristics distinctly related to a minority/ethnic group has been achieved through RSPs including dancing, drumming, rodeos, and singing (Axtmann 1999; 2001; Iverson 1999), we limit our discussion here to one ritualized practice or mode of resistance: powwow dancing (we would emphasize that we are taking a broader sociological view of Native American dancing and not one that focuses on the scholarly complexities and intricacies of Native dance as discussed by those such as Powers 1990). Through dance, "physical, social, and spiritual power is produced by the moving bodies of Native American powwow dancers as they execute specific choreographic styles" (Axtmann 2001:7). Dance provides a medium by which native peoples can reaffirm their own ethnic understandings of the physical and spiritual world. It creates a forum for Indians not only to express themselves in physical action but to also "reaffirm their ties to a living culture" (Heth 1992:8).

One of the ironies of this form of resistance, however, is that powwows and the dancing contained in them involve a blending of Indian culture, blurring the boundaries between different American Indian tribes (Bennett 1991; Mathews 1999). While American Indians gain a sense of renewed pride in their cultural heritage they oftentimes also give up distinct rituals that belong to their particular tribes. This is illustrated by the evolution of the War Dance, which is often performed at many powwows today in Oklahoma (Heth 1992:109). This dance originated centuries ago in all-male warrior society ceremonies among the Kansa (Kaw), Omaha, Ponca, and Pawnee. In the nineteenth century it spread to the Northern Plains tribes where it was called either the "Omaha Dance" or the "Grass Dance" (p. 109). In the early 1890s the tribes of western Oklahoma became familiar with it and called it the "Crow Dance." They blended it with another dance learned about the same time from the Paiute of Nevada, the "Ghost Dance." As a religious and revitalization movement, the 1890s Ghost Dance promised a return to a pre-European existence and was a "dream-like" ritual for participants to "re-create the activities of the old days." When people stopped dancing it in 1893, its ceremonial ritual centering on a large drum (the "Dream" or "Drum Dance") was dispersed into different versions of the

Omaha/Grass/Crow Dance performed by the Osage, Omaha, Ponca, and Pawnee of Oklahoma and Nebraska (p. 111).

One version of the Omaha/Grass/Crow Dance came to be called the Straight Dance. It retained much of the original prairie peoples' costume (cloth shirt and leggings decorated with ribbon-work appliqué, cross-beaded bandoliers, a headdress of deer-hair roach, and an eagle-feather) with little enhancement to its initial basic toe-heel step choreography. However, as this dance came to be performed among tribes in western Oklahoma it developed into what is now called the "Fancy Dance." Gradually its costumes became much more individually embellished, and all tribal members, including women, participated in the practice (originally it was performed only by men's societies). Its choreography was also embellished to include acrobatic spins, turns, and splits. Over time, the Fancy Dance tradition has developed in different directions to incorporate increasingly intricate footwork as well as innovations in costume design, including bustles, headdresses, and other trappings.

We see how, in the synthesis of these dance forms, RSPs are used to reproduce Indian belief systems, giving continuity to tradition by re-creating old styles of dancing and costuming while at the same time "establishing an emotional connection with more generic values of 'Indianness'" (p. 112). Powwow dancing provides an opportunity to display distinct cultural elements while at the same time emphasizing group identity and cohesion. And it provides an opportunity for the realization and affirmation of different dimensions of social power. As Axtmann states:

> [p]ower is generated when Native Americans affirm their identity as individuals, communities, and tribal nations. As Native groups unite, internal power confronts hostile outsiders.... Generational ties transfer power from son to father to grandfather to great-grandfather while the ongoing proliferation of powwows across time, geographical regions, and within diverse populations also stimulates another process of powermaking. (2001:8)

The power contained in powwow dancing is nonhierarchical and reciprocal (Allen 1986), and quintessentially "Indian"; it rejects the mind-body hierarchical dualisms of Western ideology and belief systems. Dancing in powwows is usually circular, with the drums in the middle (southern style) providing an environment in which Indians can "actively assert a specifically Indian performativity" using distinct choreographic styles (Allen 1986:9).[1] The aesthetics of Indian dancing "emphasize repetition and unison" accompanied by drumming and singing that might sound simple to the untrained eyes and ears (Mattern 1999:133). However, the dancing, drumming, and singing at powwows have a spiritual aesthetic and power that reinforce the dedication to harmony that permeates Indian culture and values, linking the individual to the tribe and the tribe to other tribes. It is not surprising then that the powwows observed in this study began with the emcee inviting everyone to join the circle, Indian and non-Indian alike. Underlying the powwow experience are the ideas of unity and inclusiveness.

Scholars argue that contemporary powwows serve as a mechanism both to transmit customs to the next generation and at the same time to maintain social and religious ties within and between Indian communities (Young 1981; Axtmann 1999; Mathews 1999; Lui 2000). In this regard, contemporary powwows can be traced historically to tribal gatherings among the Great Plains Indians when different bands would meet annually during hunts to exchange news, dances, and songs. Fearing their potential to also be used as war councils, the federal government forbade these annual meetings by the end of the nineteenth century. As part of the overall effort to colonize American Indians, bodily practices were subjected to radical control when the Office of Indian Affairs outlawed religious dancing in 1904 (Farnell 1999). For almost thirty years, Native Indian dancing gradually devolved into a popular entertainment performed primarily for white tourists (Friar and Friar 1972; Brascope 1994; Mathews 1999).

Since 1934, when powwow dancing was made "legal" again, despite governmental and racial oppression, it has thrived as a social and religious activity. It is, therefore, not surprising that contemporary powwows have become an important "cultural anchor" for urban American Indians who increasingly find themselves "homesick for Indian traditions," providing a communal connection to their past (Fixico 2000). Indeed, the large migration of Indians to urban areas after World War II prompted the emergence of numerous pan-Indian cultural practices and institutions ranging from Indian community centers to Indian Christian churches, bowling leagues, and softball teams (Hertzberg 1971; Steele 1973; Whitehorse 1988; Weibel-Orlando 1991).

We will focus here on powwow dancing because it is quite likely the most visible and dominant activity for American Indians, combining sacred traditions and practices common to many different tribes. These cultural, or to be more precise, ritual elements from different tribes are merged together to create a broader, more "generic," and all-inclusive "Indian" RSP. Such an important or dominant collectively shared ritual practice has come to play a crucial role in the construction of the distinctive collective ethnic identity of contemporary American Indians.

Though contemporary powwows vary in size from a handful of people to several thousand participants and are usually conducted by specific tribes or alliances of tribes, they are all participatory in nature and have similar ritualized dancing activities accompanied by singers and drummers. Powwows normally last from two to three days, beginning on either a Thursday or Friday and ending on Sunday. In addition to dancing, contemporary powwows can include parades, Indian Princess pageants, fund-raising events, and concession stands with a variety of native food, artwork, and crafts. Larger powwows also incorporate hand games and rodeos. All powwows begin with a formal procession, called the Grand Entry, usually led by a color guard followed by honored guests and Indian princesses. The dance competitions are interspersed with a variety of musical, dance, and storytelling performances as well as special honorings and giveaways. Indian dancing at such events is circular, evoking core Indian values of the unity between the cycle of life, the absence of linear time as either past or future, and the directional pattern of the sun (Momaday 1975).[2]

It is important to recognize that ritual practices are dynamic and subject to change. Though powwows celebrate tribal and supratribal unity, they also are forums of dissent, including disagreements and conflicts over the limited participation of women, the tension between secular and spiritual interests, and the associations between Indians and non-Indians (Smith 2000:75; Mattern 1999:127). For example, dancers are usually grouped by sex and age. Women rarely participate in drum groups and their attempts to form all-women groups have met with limited success (Mattern 1999:137). In the powwow dancing observed for this study, women dancers rivaled men's movements in agility and force, particularly in the fancy shawl dancing. Moreover, Mattern (1999) tells us that in general women dancers have been able to overcome traditional restraints that limited women's movements to being sedate and decorous. However, powwow practices remain gendered, continuing to reinforce existing power differences between men and women (Vennum 1989).

Indians also continuously negotiate the amount of weight given to secular interests over spiritual interests at contemporary powwows. As powwows grow in popularity, there is increasing concern that economic interests are beginning to drive participation in powwows. Competition for prize money in dancing and singing is starting to "professionalize" powwows, sometimes eclipsing the importance of traditional social communalism and amateur participation. Since the 1980s, more powwows have been held in urban areas than on reservations. This shift in location has been accompanied by an increased emphasis on economic profit sometimes at the expense of social or religious renewal (Mathews 1999; Peterson 1999). Larger powwows offer substantial prize money for their dance competitions, which has resulted in a proliferation of "professional" powwow dancers who now can support themselves by traveling on the powwow circuit. For instance, the annual Black Hills Powwow in Rapid City, South Dakota, awards almost forty thousand dollars each year. And, tribal-based, casino-sponsored powwows such as the one held each year in Hinkley, Minnesota, have resulted in intense competition among contestants, diminishing the previously communal nature of such events (Mathews 1999:72).

Powwows have also sometimes been appropriated as a political forum to focus attention on pan-Indian grievances. For example, members of the American Indian Movement (AIM) have organized powwows to protest the use of American Indian names and mascots by athletic teams (Mattern 1999:139). At the same time, some Indians think it is inappropriate to use powwow rituals to pursue secular economic or political goals.

Another concern is the growing involvement of non-Indians at powwows. Though non-Indians are welcome to participate, this invitation is extended as long as they are willing to respect indigenous cultural practices. However, it is more and more a common sight to see non-Indians inappropriately improvising while dancing or photographing dancers and drummers, disrupting the flow of action. Powwows provide a microcosm of how Indian relations with non-Indians are negotiated in the broader society, either through choosing to disregard inappropriate behavior, mocking it, or engaging in outright intervention (Mattern 1999:138).

Notwithstanding these developments, powwows have continued to be a profoundly important ethnic experience for many American Indians. It is a collective experience that to varying degrees still unites all (or most) actors in a common ritualized Indian practice. At the same time it provides a way of signifying to others in the wider society the existence and efficacy of this ethnic group. Indeed, today American Indian tribes have more autonomy and are more self-sufficient than at any time since the last century (Snipp 1996). Despite the disease, warfare, and famine that ravaged their ranks by the end of the nineteenth century, American

Indians have been able to resist extinction and acculturation and to varying degrees improve their political and economic status.

Some attribute this remarkable resurgence in autonomy and self-sufficiency to the mobilization of modern pan-Indian organizations in urban areas (Cornell 1988; Fixico 2000; Nagel 1982). Probably the most compelling argument is put forth by Nagel (1997), who suggests that this resurgence of American Indian autonomy is due to the recent pan-Indian ethnic renewal that has shaped the American Indian ethnic identity of both the individual and collective selves. Nagel argues that at the individual level, ethnic renewal is achieved through "the acquisition or assertion of a new ethnic identity by replacing a discarded identity, adding to an existing ethnic identity repertoire, or filling in a personal ethnic void" (Nagel 1997:9-10). Examples of individualized ethnic renewal would include participating in a traditional American Indian spiritual ritual such as a sweat-bath or performing as a dancer, singer, or drummer in a powwow. At the collective level, American Indian ethnic renewal involves the reconstruction of community, that is, old or new members building or rebuilding community institutions by drawing on Indian culture, traditions, or history.

Though Nagel (1997:11) points out that large numbers of individual and collective ethnic renewals "do occur," she does not fully articulate how these kinds of ethnic renewals actually come about. Expanding on the insights of Nagel, we suggest that collective ethnic renewal is most effective when it involves the creation and strategic use of a program of new or revised RSPs such as powwow dancing. Powwow dancing provides a glimpse into how collective ethnic renewals that are created at the individual level are simultaneously created at the collective level as well. In this case, a ritualized practice realized through dance is a dynamically embodied action that is engaged in resistance to acculturation at the same time that it is sustaining a distinct Indian ethnicity. As an individual RSP, powwow dancing becomes an embodiment of human agency: by using the body, as lived space, it is an important site of action, gratification, meaning, and resistance in collective ethnic renewal. In doing so, powwow dancing is also an important ritualized practice related to the creation of a collective-level pan-Indian ethnic identity. It is one way in which American Indians have been able to fashion an autonomous cultural identity in contemporary society. By producing and engaging in a program of ritualized collective events called powwows, American Indians have been able to create a pan-Indian identity based on a syncretic conception of "Indianness."

In sum, like the Nazis and the Orange Order, many Native American participants strategically use ritualized practices in collective events to achieve certain ends. For Native Americans, their goals include sharing an enjoyable and meaningful experience, the creation of a sense of social self—that is, a collective Indian ethnic identity—engaging in a religious activity, and establishing a greater sense of efficacy as a viable social entity. In doing so, Native Americans actively foster the reproduction of their past while at the same time resisting assimilation into Anglo society, thereby affirming the power of their group and the sense of empowerment that flows from such acts. Powwow dancing is a key ingredient in a program of RSPs employed by members of this group.

Such a program, however, compared to the other two under consideration is much more decentralized (i.e., tribally based), less formally institutionalized, and less clear-cut in terms of explicit and shared reasons for production and enforcement. The impetus for the formal production and support of powwow dancing emanates from a variety of individuals and groups, including those who are driven by the desire to express and share with others their "Indian" identity and culture, those who derive personal and social gratification and meaning from participating, those who see such activities as a form of political exercise and statement, and those who profit economically from their enactment. The construction of this ritual program is often the product of various parties, driven by their respective objectives and negotiating and working with one another to create such ritualized events.

Comparison and Analysis

Three different cases of ritual/power dynamics have been examined in this study. The first case, Nazi Germany, provides us with an example of an extensive and quite comprehensive ritual program comprising a wide array of RSPs that to a large degree complemented and reinforced one another in communicating certain messages to all members of society. This program was the product of a state regime that was able to implement and enforce its production at nearly all levels (i.e., micro to macro) and in most institutional sectors of society. In the case of the Orange Order, while other types of ritual activities exist to varying degrees, we find that one particular type of ritual practice stands out, that is, parades. A highly organized political-religious group, the Orange Order has quite effectively conducted this type of ritual program for many years. It has enforced this ritual enactment throughout an entire society while preventing contending groups from engaging in the same practice. In the third case, a ritual program has emerged in recent decades among Native Americans primarily composed of the ritualized practice of powwows along with other less dramatic ritualized activities. Here, different parties pursue various ends, one of the most important ones being the reaffirmation of their ethnic identity, as this ritual event is enacted. It is the product of various groups essentially from within the Native American community working together. As such, it is less institutionalized and complex in its overall form and design.

Whatever the nature and form of the ritual program employed by the particular group, we find that in all three cases the use of rituals has been of crucial importance to the group. Furthermore, we find that the organizational/power structure of these groups has played in different ways a key role in the production and enforcement of ritualized practices. Finally, although the specific goals pursued by these different groups have varied, they have often (although not exclusively) been aimed at protecting, if not enhancing, the power of the group and/or fostering a sense of collective identity.

It is important to note that the close examination of these three cases reveals that ritual programs exhibit great variations. In comparing just these three cases we find that programs of ritual practices differ in terms of the types of ritualized symbolic activities engaged in, the number of times rituals are enacted, and the overall complexity of the program. We therefore suggest that ritual programs may be placed along a spectrum with those at one end exhibiting the greatest complexity (i.e., most types of ritualized practices, highest frequency of ritual enactments, greatest degree to which different types of RSPs complement one another and are similar in terms of the symbolic themes expressed by the activities) and those at the other end of the spectrum exhibiting the least complexity.

In regard to the three cases examined here, the Nazi party would be located at the most complex end of the scale, Native Americans would be at the opposite end of the spectrum, and the Orange Order in the middle. Of course, these are only approximate assignments. A more in-depth study of any of these three or any other groups could reveal further details that would allow for a more nuanced and fine-tuned assessment of *ritual program complexity*.

Examination of just these three cases also reveals that great variations can exist among ritual enforcers, that is, the ways groups strategically engage in this type of ritual enactment. Here, we have directed our attention to ritual enforcement by a totalitarian state regime, a political-religious group within a society, and ethnic groups and communities. In these cases we find clear differences in terms of the type of social system conducting and enforcing ritualized practices; the degree of coordination and organization within the group that enables it to produce these rituals; the extent to which such activities are institutionalized or formally codified, assigned, and carried out by group members; and the extent to which ritual activities result from the actions of a single interconnected, highly regulated social entity or are the product of different parties and subgroups that are either working together, negotiating with one another, or are in contention.

Such variations in ritual enforcement could again be identified by locating groups on a spectrum or at points of a gradation signifying greater or lesser degrees of complexity in regard to these attributes. We therefore refer to this concept as *ritual enforcement complexity*. In this context, the Nazi party would as before be at the highest end, Native Americans at the other end, and the Orange Order in the middle. Though representing an approximate categorizing of these groups, we believe this preliminary assessment demonstrates the potential value of this conceptual device that could be further refined and used in subsequent research and analyses of strategic ritual enforcement.

At the same time, this investigation raises other questions and research issues. For instance, more detailed study is needed identifying additional, perhaps more subtle or smaller, ritualized practices that may or may not complement those that have been identified here (including possible practices in groups where one particular type of activity appears to be dominant). Moreover, greater attention could be given to the organizational makeup of these groups. Further research could examine in greater depth the internal organizational dynamics operating in these social units that may positively or negatively affect ritual production. Finally, an important question that warrants attention deals with the impact of these practices on actors: To what degree are people actually influenced by these ritualized practices? In that regard, there is a need to delineate and measure the processes and conditions involved with ritual enactments that contribute to their potential impact on people. All of this work could involve the extension and development of concepts within SRT.

In conclusion, these concerns and all of the work described in this chapter rest on certain presumptions and arguments. These arguments and the findings of the studies discussed here are that ritualized activities are oftentimes strategically engaged in and done so in different ways involving strategic legitimators, strategic entrepreneurs, and strategic sponsors; another type of strategic ritualization involves rituals and social power, that is, ritual enforcement; ritual enforcement occurs throughout the world and through history; ritual enforcement in different cultures and periods is evident in the three cases examined here; ritual enforcers may differ in a number of respects, particularly in regard to organizational and social attributes, that is, ritual enforcement complexity; groups may employ ritual programs or collections of RSPs to strategically achieve certain ends; and, ritual programs can greatly vary in terms of ritual types engaged in, frequency of ritual enactment, similarity of ritual types, overall structure, and other possible features, that is, ritual program complexity.

Finally, having conceptually identified and provided empirical evidence for this phenomenon in this cross-cultural study, future studies will examine more systematically and in greater depth the dynamics by which power, rituals, their enforcement, and their consequences are realized in different settings. In doing so, such work should not only further our knowledge of these specific issues but, more generally, increase our understanding of how ritual enactments and culture are integrally involved in the structuring of interaction, organizations, and society, thus providing a missing link in sociological analysis.

Notes

1. There are two major powwow styles: northern and southern. The basic difference is the placement of the drums, which are outside the circle in the northern style and inside in the southern style, as in Oklahoma (see Champagne 1999:127).

2. Powwow dancing is also a ritualized practice that can be viewed as a "conversation" involving mannerisms, gestures, and postures. The dancers, drummers, and singers all dress in a variety of Native costumes, usually made by the participants themselves. Some combine denim with traditional tribal garb while others exclusively use buckskin adorned with beading and feathers, sometimes embellishing their costumes with different types of cloth or bells. This extensive mixing of different tribal attire, dances, songs, and customs actually makes some American Indians uneasy. Though all tribes are invited to join, some tribes, such as the Southwest Pueblo stay away (non-Pueblo Southwest tribes such as the Navajo, however, actively participate) (Lui 2000).

3. Using Bourdieu's (1977) notions of "habitus" and "nexus" to expand

generally on the role of habitual body and spatial practices in social action, we can begin to theorize how the body and body movements can be specifically studied as a form of ritualized communication. In general, the body is largely absent from sociological discourse, having been marginalized from social interaction by being labeled as individual, psychological, or sensual response (Farnell 1999). This research, however, hopes to recapture human movement as social action that is embodied and significant. Once body movement can be conceived of as a wordless social act, we can begin to see it as empirically situated and relevant, rather than dismissing it as an instinctual response. Powwow dancing is body movement, which as an RSP models social relationships, and symbolizes ethical and political commitments. Seen this way, powwow dancing takes on broader political, economic, and social implications and is of direct relevance for the power of groups and individuals.

Chapter 5

Disruption and Deritualization

Concentration Camp Internment and the Breakdown of Social Order

Frédérique Van de Poel-Knottnerus
and J. David Knottnerus

This chapter addresses a different line of research that focuses on the basic assumption underlying SRT that rituals are a crucial and necessary dimension of human behavior. This perspective is grounded in the basic assumption that daily life is normally characterized by an array of social and personal rituals. Such rituals, we suggest, help create stability in social life while expressing various symbolic meanings that give significance to our actions. We address this issue by examining situations in which the everyday ritualized practices people normally engage in—those that make up their taken-for-granted life—are severely disrupted, and the ways people can cope with such occurrences.

We assume that when patterned, ritualized modes of behavior are severely disrupted, this is a very difficult and problematic situation for human beings. It is in such situations that the importance of reconstructing RSPs is most evident. If people can reconstruct old or new rituals after such disruptions, this enables individuals to reorganize their daily, taken-for-granted lives. Such actions create newly structured social and personal patterns grounded in ritualized behaviors and relations. By creating a new system or pattern of rituals people achieve a meaningful focus, sense of direction, and stability in their everyday lives. In providing meaning, direction, and coherence to our actions they at the same time direct our minds away from conditions that may, among other things, be incomprehensible or chaotic, detrimental, aversive, distasteful, tedious, or boring.

Stated somewhat differently, individuals and groups who are able to create new or reconstitute old ritualized practices subsequent to such disruptions are best able to cope with and adapt to their immediate situation. These collective and more "individualized" rituals (which are still social in nature) serve as a buffer to disruptive events. Research that addresses this basic issue provides evidence for the presumably universal occurrence and importance of rituals for providing meaning, significance, and a sense of order in people's lives.

Given the validity of these assumptions we expect that the more coercive or extreme the disruption the more apparent will be the harmful effects of such an experience. While limited attention has been given to this issue we would point out that Giddens (1984:61; see also Garfinkel 1967; Maynard 1996; Berger and Luckmann 1966; Goffman 1961; 1967) has discussed a related concept referred to as "critical situations," which he defines as "circumstances of radical disjuncture of an unpredictable kind which affect substantial numbers of individuals, situations that threaten or destroy the certitudes of institutionalized routines." While his approach rests on the work of Erik Erikson and the concept of ontological security, his analysis and the argument presented in the present book are to some degree similar and are relevant to several situations ranging from imprisonment in concentration camps to "being under fire on the battlefield for prolonged periods of time, forced interrogation and torture in prisons and other conditions of extreme stress" (p. 64). We suggest that individuals who successfully cope with or adapt to such critical or extreme disruptions of their everyday ritualized behaviors are those who are capable of reestablishing RSPs in their personal and social lives.

Disruption, Deritualization and Reritualization

More precisely, SRT addresses this topic by examining what are referred to as disruption and deritualization and the ways different social groups and individuals respond to such experiences. *Disruption* (or disruptive events, conditions, and developments) refers to occurrences that disturb or interrupt the RSPs that individuals and groups normally engage in. Numerous kinds of events we assume can lead to the breakdown of ritualized practices, relations, and social structures. Such events can vary in their intensity and can profoundly impact the daily lives of people, resulting in differing degrees of deritualization. *Deritualization* involves the breakdown or loss of previously engaged-in RSPs by groups and individuals. It refers to the breakdown, collapse, or cessation of ritualized activities that occur in daily life.

It is worth emphasizing at this point that SRT stresses that ritualized practices contain two key components (both points of which reflect the basic definition of RSPs as schema-driven action repertoires). RSPs involve both human action and cognitive or symbolic frameworks that possess meaning. These two elements provide the fundamental parts of RSPs and the two basic dimensions of a ritualized practice, which are subject to breakdown during deritualization. Deritualization is realized through its effects on meaning and behavior in rituals. There is a loss or breakdown of both action and meaning in people's everyday lives. At the same time, although both are usually involved in deritualization, the extent to which they are present can differ. For this reason, one can conceptualize deritualization along these two dimensions, thereby making it possible to approximately assess whether such experiences are more representative of breakdown of action, the loss of meaning, or both to roughly the same degree. Ultimately, however, deritualization to varying degrees always involves both critical dimensions.

A crucial way people can cope with disruptions and deritualization is to reengage in ritual enactments. Such enactments serve as buffers that enable actors to manage the harmful consequences of deritualization. *Reritualization* refers to the re-creation of RSPs following disruption and deritualization. By reconstructing RSPs subsequent to disruption and deritualization, people can reestablish essential behavioral repertoires of a personal and social nature, which enable them to more effectively deal with their situation. Reritualization makes it possible for actors to re-create meaning and a focus in their actions. These renewed ritual enactments re-create coherence in actors' perceptions and actions; they generate a cognitive clarity and plan of action, which create a fundamental sense of stability. As individuals become regrounded in the taken-for-granted world, they acquire a sense of reassurance and security.

These kinds of phenomena, while quite varied in nature, are, we suggest, quite common. As such they deserve greater scrutiny. Upon closer reflection, however, one quickly realizes that there are many unanswered questions surrounding disruption, deritualization, and reritualization in human behavior and social life more generally. For instance, do disruptions, especially negative ones, differ in both their nature and degree of severity? And, when people experience disruptions, especially severe ones, what are the exact nature and consequences of the experience of deritualization? Do people engage in reritualization after a disruption? And if they do so does the reconstitution of RSPs actually enable actors to cope with such events? And what kinds of rituals are involved in reritualization in different types of situations? Furthermore, one could ask whether ritual enactments and the harmful effects of deritualization are so basic to the human condition that we find such processes occurring in very different social settings, groups, and cultures. One line of research in SRT has begun to address these issues.

Nazi Concentration Camp Internment

To address such concerns, this line of research examines the "personal accounts" of individuals interned in concentration camps to study extreme instances of deritualization caused by coercive interventions, and whether the reconstruction of ritualized practices enables actors to cope with such disruptions (the next chapter highlights subsequent studies of other groups). The first investigation of this topic examines one type of deritualization, referred to as "institutional deritualization," which involves the forced entry of actors into total institutions where they experience an extreme, coercive disruption and breakdown of previously engaged in RSPs occurring in numerous domains of interaction in their normal social life prior to that disruption. The first study of this issue (Knottnerus 2002) is a comparative case study focused on persons placed in concentration camps in the middle of the twentieth century in four different cultures: Germany, Japan, the Soviet Union, and the United States. Evidence about individuals' experiences in these quite different cases was obtained from a number of sources, especially published accounts by survivors. Twenty-six accounts published in English were investigated, spread across the four cultures (five to seven accounts for each group were examined).

More recent research—to be described in this chapter—expands on this first study by examining in a more focused and in-depth manner the experiences of far more persons interned in Nazi concentration camps. And we direct our attention to the experiences of French and non-French prisoners, both female and male, and mostly adults (although a few are teenagers). This collection of accounts provides a culturally and linguistically diverse sample for analysis.

Evidence concerning actors' experiences in these different cases has been obtained from published accounts in the form of memoirs, autobiographies, diaries, interviews with prisoners, biographies, and certain books and essays by former internees, dealing with related issues such as the psychological dimensions of imprisonment. The accounts deal with both French and non-French internees and have been published in either French or English. We have examined seventeen French and twenty-eight non-French accounts for a total of forty-five accounts. These sources were obtained through the use of university library facilities and research conducted by the authors of archival documents at the United States Holocaust Memorial Museum in Washington, DC (supported by a grant to Van de Poel-Knottnerus from the College of Arts and Sciences, Oklahoma State University). All of the accounts in French were translated by Van de Poel-Knottnerus. We list all of the sources categorically for French and non-French authors in a separate subdivision at the end of the reference section. Most accounts were produced and distributed by publishing companies of varying sizes, although several are either self-published or produced by organizations or academic institutions concerned with the Holocaust. A number of these accounts including self-published sources were found at the U.S. Holocaust Memorial Museum.

In reading, recording, and analyzing these narratives, we have paid special attention to and have collected evidence concerning a number of topics dealing with the experiences of persons interned in these camps and the ritual dynamics operating in these settings. While accounts were randomly selected, one essential criterion is that they must contain narratives that describe the personal and social experiences and behaviors of prisoners. If they do not, it is not possible to derive any information about the ritual dynamics that potentially occur in these settings. While most accounts do contain information about people's behavior, occasionally they do not. This is due to various reasons, including the shortness of some documents, the writer simply not addressing this issue, or authors focusing on other concerns such as only describing their captors and the structure of the camp, simply chronicling their movements from one facility to another, or only discussing philosophical, theological, or moral questions.

In regard to the accounts examined we compiled the following evidence. First, descriptive information about the type of camp and conditions within it, to the degree it is available, was collected. In regard to the experiences or behaviors of internees one item of information that we have recorded deals with any references to the effects on prisoners of forced placement into a camp, that is, the condition of deritualization caused by this disruptive event. Here special attention is given to how these events impacted actors' ability to meaningfully interpret the world and engage in coherent actions.

By reading these accounts and employing certain rules and a coding scheme, we also identified different ritualized behaviors that were sometimes engaged in by prisoners as a response to these severe disruptions of their lives. Briefly stated, the rules guiding our study and the first investigation are as follows. References to a ritualized practice must be clear, that is, unambiguous. The account should clearly indicate the type of activity, that is, the content of the ritual. And it should be clear who is engaging in the RSP. To be sure that all ritualized practices are identified and not just practices that might seem "obvious" to a researcher (such as a birthday celebration), it is necessary that practices correspond to definitional criteria. They must involve action repertoires, the acts must be standardized or regularly engaged in, and they must express or communicate certain symbolic themes or meanings.

The coding scheme for identifying direct references to RSPs that internees engaged in was developed in the previously mentioned cross-cultural study. It involves categorical distinctions that identify substantive differences in ritual practices, that is, ritual contents. The collection of categories indicating different ritual contents includes but is not limited to religious rituals; special lectures and presentations; political acts; classes for adults; talking, storytelling, and discussion groups; holiday celebrations; sports; games (e.g., cards, gambling); concerts, shows, and theatrical productions; birthday celebrations; dance; music; parties;

artwork; crafts; reading; schooling for children; food preparation, eating, and recipes; gift giving; walks and hiking; gardening and outdoor work; writing (e.g., newspapers, newsletters, literature); diary keeping; special ceremonies and events (graduation, cultural festival, celebration for some type of accomplishment); recreational activities other than sports and games (e.g., astronomy); medical treatments; and idiosyncratic activities (unique ritualized activities peculiar to an individual or group). A total of twenty-seven content types are identified in this first collection of ritual categories.

When a practice falls into more than one category usually a determina tion is made as to which type is most relevant. However, it is sometimes necessary to identify several categories for a particular act due to their importance, for example, a special holiday celebration that is religious in nature. Information of a more detailed nature has also been collected about consistencies or differences of rituals within categorical types. Finally, as research continues to address this topic, other categorical distinctions may be identified, while more limited categorical schemes may also be developed that group various rituals within more general or abstract ritual types (as has happened in recent research examining other forms of disruption and deritualization, which will be discussed in chapter 6).

Looked at from a different angle one also finds that these activities take one of three forms and can be interpreted as such: quasi-formal organizational rituals, informal social rituals, and personal/individualized ritual practices. Ritualized practices may be more overtly collective or social in nature and exhibit to varying degrees an organized and more formalized quality. On the other hand, RSPs may be carried out in a more informal, less regulated, less structured manner. Lastly, people may engage in ritualized actions in a more private manner by themselves. Examples of these could involve, respectively, prisoners developing and participating in a sports club and tournaments, repeated interaction between internees as they share recipes or memories, and individuals writing diaries.

In addition to direct references to specific RSPs, another type of evidence has been gathered. We have given special attention to any type of interpretative or reflective statements by actors about ritualized practices engaged in by themselves or others, the importance of RSPs for survival and coping with conditions in the camp, what persons were better able to adapt and survive and the rituals they engaged in, and individuals who were incapable of coping with camp conditions.

An assessment is also made about the survival of internees. Several types of evidence make it possible to determine with a high degree of confidence whether a prisoner was able to adapt to camp conditions and ultimately survive. Usually information is provided in accounts about this issue. In most cases these accounts were completed and provided by individuals after their release so we know they were among those who survived imprisonment. It is also usually possible to assess from these accounts how the individual responded to the environment. Indeed, in a number of cases very explicit commentary is provided about who was best able to survive, the reasons they did, and the importance of different practices and social conditions for internees' strength and stability.

Finally, we have noted any references to ritualized activities engaged in by actors immediately after their release from the camps, that is, "post-internment ritualized practices." While potentially useful, we have found that this type of commentary is only occasionally provided in sources.

Various findings have emerged from our research into the experiences of French and non-French internees in Nazi concentration camps. These findings are of a qualitative nature. The first study of concentration camp internment contained more quantitative evidence concerning, for instance, approximate number of references to ritual practices in accounts. In the present study our emphasis is on what qualitative evidence, in the form of narratives and actual observations by a much larger number of internees, reveals about their experiences in Nazi camps.

First of all, we find that descriptive information about prisoners such as gender, age, nationality, preinternment status, and occupation is normally provided in accounts. And conditions in camps and their organization are usually described, oftentimes in quite vivid terms. Moreover, definitive information about the survival of the internee is almost always provided. Based on this information, our assessment is that nearly all of the internees whose accounts we studied survived and were able to at least a minimal degree cope with the highly disruptive, coercive placement into camps and the breakdown of the ritualized practices that defined their taken-for-granted daily lives prior to internment.

Turning to the actual experiences of these individuals, we find that in a number of accounts statements are made concerning the debilitating effect that placement in a concentration camp has on actors, that is, the effect of disruption and deritualization. A little less than half of the narratives we have examined contain some sort of description of individuals' difficulties or inability to comprehend the world they are in and/or to act in a coherent, focused manner. Some of these statements are quite revealing and poignant. Indeed, we find that the words of these survivors provide in many cases the most compelling evidence for the analytical arguments presented in this chapter. For this reason we will illustrate this and subsequent points to follow with a sampling of narratives provided by both French and non-French prisoners.

For instance, consider the words of Mechanicus (1964:129): "Anyone who cannot bear his own suffering and that of the masses becomes slowly or rapidly submerged. Cases of mental disturbance occur in men and women who in normal society were known to be sturdy, resilient folk; they have had a shock." A more personal description of the experience is provided by Wiesel (1982:33–34) who says: "We were incapable of thinking of anything at all. Our senses were blunted; everything was blurred as in a fog. It was no longer possible to grasp anything ... we were damned souls wandering in the half-world, souls condemned to wander through space.... Within seconds, we had ceased to be men.... So much had happened within such a few hours that I had lost all sense of time.... Surely it was a dream."

Others also employ similar or related imagery in describing the loss of all that was taken for granted and normal. Perl, a doctor in Auschwitz (1984:46–47), refers to the "ghostly carnival" and beautiful women looking "like a bunch of gristly monsters, ridiculous, and sub-human." So too, Weiss (1961:42–43) portrays the changes occurring in the prisoners as "ridiculously grotesque" stating that "Everything whirled around me.... Is this not a ghastly dream?" Then, she pointedly notes: "What had happened to us was completely beyond understanding."

Further commentary expresses the same themes. Newman (1963:20) employs extraordinary language to depict the altered, alien condition she feels herself to be in: "I found it hard to believe that we were still in Europe, and sometimes I felt I had been carried away to some primitive island where I was forced to live like a savage. Many times I felt I must be dreaming, and I would call to myself: 'Wake up! Wake up! You are having a nightmare!' ... my eyes kept on seeing the same dismal picture." She concludes with the observation: "Never had I read about anything so horrible as what I was seeing with my own eyes and living through myself." So too, Alizon (1996:114–16) graphically describes the impact of internment and sense of loss induced by this event: "The first room we entered marked the beginning of our transformation ... To be stripped of a part of ourselves in such little time ... Such a shock. Such a profound shock! What can be said about the intense feeling of fragility, of vulnerability, felt by all who experienced such a degradation of their being, so total, so fast, so brutal. Such a shock. Such an abominable shock. Such a profound offense done to each of us."

Nomberg-Przytyk (1985:15) also captures the sense of how this extremely disruptive experience resulted in the incomprehensible loss of all those familiar practices and customs that had constituted her previous life: "In a few hours we were robbed of everything that had been ours personally.... Nothing was real. It was as if I were looking at a picture from another world." She then goes on to state: "From the moment I got to Auschwitz I was completely detached. I disconnected my heart and intellect in an act of self-defense, despair, and hopelessness" (p. 18). Others discuss in similar terms how their lives have been profoundly changed. In this regard Holt (1995:25) states: "As soon as I cross the gate, I have the acute feeling of leaving one world for another one. On the other hand, what I cannot predict is how much this new world will involve the total distortion of the one I just left." Rovan (1993:40) describes how he finally began to mentally readjust, after the profoundly disruptive forced placement into the camp: "I came back to my senses. I did not feel as distressed as when for instance something irrevocable has happened, when an irreversible event places us in a new world."

The impact of this disturbing event on people and the resulting breakdown of their world and the ritualized behaviors, cognitions, and relations that define individuals' daily lives are also described in other ways, including more analytical language, as in the case of Cohen (1973:46) who discusses how he underwent a "'depersonalization.' You looked on, as though through a peephole, taking no part in things yourself. You watched. It didn't concern you." So too, Szmaglewska (1947:78) states: "Your feeling of individuality is entirely lost. It seems that ... you lost ... the visible sign of your identity. You lost the capability of proving to yourself ... that you are still the same human being you were when you came here. That being is gone, and only a miserably wretched creature remains in her place." Later she goes on to describe how the camp [Birkenau] became a "jungle where you may easily go astray.... The husks of principles fall away and so do the trite forms of good behavior.... Man is left naked and deprived of his buckler ... [some] like an amoeba, dissolve in the general moral chaos of the camp" (pp. 284–85).

In regard to the possibility that people may engage in reritualization, that is, create or reconstitute ritualized behaviors in such extreme circumstances, we find that all of the accounts (forty-five out of forty-five) that were authored by survivors contain explicit references to RSPs enacted by these individuals. Every narrative contains statements about rituals practiced by the individual and sometimes by other internees. While a wide variation exists in how often such references are made by different authors the overall extent to which such references occur in the entire collection of accounts is quite striking. Moreover, comments refer to a wide variety of ritual types or contents. In looking at the entire collection of accounts, many of the twenty-seven different types of ritual practices identified in previous cross-cultural research are mentioned by internees, including, for example, religious practices; talking, storytelling, and discussion groups; holiday celebrations; music; sports; and idiosyncratic activities.

Certain ritual contents appear to be more common, such as references to religious or musical activities. We would stress, however, that observations such as these can be highly misleading because a reading of different accounts reveals great variations. For instance, while religious practices may be cited numerous times in one source, in another account they may not even be mentioned or receive less attention than other ritualized activities such as celebrating holidays, producing shows, playing games, or talking and storytelling.

An examination of the actual comments of a few of the internees provides a graphic illustration of these observations. For instance, Herzberg (1997:115–16) informs us that because religious practices were forbidden in the camp, "religious observance took place in *every* hut, despite the work that still had to be done that day, despite three roll-calls, despite the air-raid alerts, despite the pouring rain. We can meet before work, before six in the morning. Those who remain inside the camp also meet between roll-calls, and after half past six in the evening, when the working parties have returned, there is the opportunity for the final prayer." So too, Matzner (1994:viii–ix) states: "Often I saw men at prayer in the camps, gathered in a corner of their barracks, their bodies swaying, their eyes closed, their lips moving in prayer. I never participated in these improvised services.... Yet each morning I would say aloud to myself and to those near me, 'Today is one more day. This day, too, we shall overcome with the help of God.' Then I would recite a few lines of that beautiful prayer, Adon Olom ('Lord of the World')."

On the other hand, Beattie (1946), a war correspondent imprisoned in Nazi camps, repeatedly describes prisoners as engaging in craft and artistic activities or practices associated with food, spending hours creating a meal, constructing recipes, and taking part in cooking "bashes," which sometimes celebrated special events or holidays. Elsewhere, in a memoir written by a Polish woman (Nomberg-Przytyk 1985:114–16) who was interned in Auschwitz we are informed that a group of women prisoners "organized a cultural evening in the infirmary ... [and] planned an evening of dance ... [and that they] often organized such friendly get-togethers." Here they would, among other things, sing French and Russian songs, tell jokes, recite poems, sing prison hymns, and play Hungarian and gypsy melodies with a violin.

A different type of practice involving medical, political, and collectivist motivations is described by Leroy, Linet, and Nevers (1991:112): "In the evening, in the block, while the regular meeting was secretly taking place to give care to the many wounded, Roger Linet had gathered four or five militants to explain that he didn't think this solidarity should be restricted to our group. We were all in the same boat here; we had to help one another." Others such as Frankl (1972:55) tell us that occasionally a "scientific debate developed in the camp" and that he began to "reconstruct" in his mind and on scraps of paper a scientific manuscript that he had lost. And various other authors describe a variety of ritualized practices that they or others engaged in ranging from the staging of cabarets and chess tournaments in Westerbork, a predeportation camp (Mechanicus 1964), to engaging in repeated chats, conversations, and visits with friends and acquaintances (Kielar 1980) and staging Sunday boxing matches (Ka-Tzetnik 135633 1963).

Even in quite coercive camps activities such as these are described, for example, in the case of Rosenfeld (1991:64), who states: "Evenings of 'variety shows' were organized in some barracks. And it is not rare to find true talents. There even are 'Professionals' who sometimes are enchanting.... But there were not only songs. I already mentioned, lectures—usually always interesting—on various topics, given by different individuals. We were discovering anew parlor games: checkers, chess and even 'little horses.'"

The importance of RSPs is also apparent in those accounts that describe the extreme measures internees would sometimes take to carry out such activities. In this regard Borwicz provides the following vivid narrative:

> In order to talk about the theatre, says Cz. Ostankowiscz, we would meet in a small square near the death wall. The one used for capital executions. This was the safest location, because the Germans did not like the sight of pyramids of cadavers. It was also the most convenient because,

> after work, the garbage cart was placed there and Jaracz would use it as a podium for his art classes ... Jaracz ... asked me a question: ... You have a theatre, here? ... Yes, a theatre. We must provide one to the men because words help them to hold up their heads. There are artists and men of letters here. We must suppress any way we can the Germanic drums' sounds and trumpets' roarings. (1996:134)

Clearly, prisoners engaged in a highly diverse array of ritualized behaviors. Moreover, we would stress that wide variations exist within types of ritualized practices or ritual contents. For instance, religious practices take quite different forms. Herzberg (1997:142–44) documents how this is the case when he describes different groups celebrating the Jewish month of feasts and the last night, Rejoicing of the Law: among the North Africans, he encounters great excitement, rapture, and intense dancing; with the Balkan Jews a very different orientation is seen, with some even converted to and practicing Greek Orthodox religious customs while others are engaged in Jewish mysticism; and among another group, their religious activities are totally focused on nationalistic themes and Palestine.

Furthermore, we find that certain individuals discuss idiosyncratic ritualized activities. Take for example the following comment from a prisoner, Gill (1988:360), who survived internment in Auschwitz: "I kept to a ritual of washing—although they made water available in the Block washroom only at midnight. All through the summer of 1943 I remember getting up at midnight, which was an effort, and going to wash myself. I know that having a self-imposed routine was a tremendous help." Later, he further discusses this point by noting the significance of a different kind of ritual activity for some prisoners in the camp. While it involves religious ritualized practices it takes on a quite unique, extreme, and in this sense idiosyncratic form: "there were some—some who even fasted on Yom Kippur, some who tried to keep to a kosher diet. Maybe the ritual discipline helped, just as the ritual discipline of keeping clean was important."

Actually, the quite distinctive ritual of washing is referred to in at least one other account that we have examined. Nomberg-Przytyk (1985:19, 107) describes how her "usual custom" was to go to the infirmary before roll call because there was running water there, which she could use to wash up. She also offers the following pointed observation: "It was important to wash, even if it meant rubbing your face with a fistful of snow. The effort to wash your face is an expression of life. The women from Holland [who were imprisoned in the same barracks] did not fight to survive. They very quickly became 'mussulmen,' and many of them died [quite quickly]."

A final example of an idiosyncratic ritualized behavior is provided by Matzner, who describes one such practice he engaged in while he was interned in numerous Nazi slave labor and concentration camps:

> After the constant pressure and unbearable pushing of the kapos, the malicious watchfulness of the guards—after all this, the camp meant a temporary escape, something like home. I could crawl onto my board there.... When we marched out in the early morning I was already longing for the coming night so that I could once again stretch out on that little bit of space that, however dirty, cold, and overrun by lice, was mine. The chopped straw in the sack, the filthy blanket, two rusted tin cans, one twisted spoon, a broken fork, a few foul and stinking rags: these were my accessories, this was my "home."
>
> Each evening when we returned, my first step was to check my possessions, relieved to find everything intact. (1994:51–52)

The significance of these materials, which were essential components of (or resources for) this RSP, is clearly apparent when he is being moved to another prison and states: "I took my spoon, my toothbrush, a piece of glass from a broken mirror: These were my possessions and I would not give them up" (p. 90).

Narratives such as these show how unique and small, perhaps seemingly trivial, rituals can, under extreme conditions, assume great significance for a person, thereby providing some structure and stability in their lives. Under these circumstances such practices may take on great symbolic import for actors. Ritualized practices, as previously discussed, are based on cognitive schemas and convey and express meanings or symbolic themes. They are significant to individuals, they provide a focus for and coherence to behavior and our interpretations of reality, and they direct our minds away from conditions that may be disliked, repulsive, or not understandable.

These examples of idiosyncratic practices shed light on the value of these activities for individuals and how they believed rituals enhanced—if not were crucial to—their ability to adjust to the extreme disruption and conditions they were subjected to, and therefore increased their chances of survival. In that regard, we find that a number of accounts contain quite revealing reflective comments by internees about camp conditions, behavior, and actors' ability to cope with their environment. We estimate that approximately a third of the accounts contain some form of self-reflective statements.

For instance, in discussing camp life, Beattie notes that those

> who were mentally equipped to make the best out of their lives made a point of finding themselves some form of activity which could eat up the waking hours and prevent staleness. It's the men who lie in their bunks pitying themselves, complaining everlastingly of the food, the bedding, the weather and the war, who go first to seed and who, granted that no prisoner is entirely sane, are most apt to succumb to mental disorders. (1946:226–28)

Other commentary is less specific in its descriptive nature. Nevertheless, the statements clearly indicate the importance of regularly engaging in some type of meaningful activity. For instance Birenbaum (1996:ix) states: "I did not make notes with the idea of using them later; they simply helped me in some way to survive." Szmaglewska (1947:272–73) also states that those who survived the epidemics that spread through Auschwitz and Birkenau began "to adjust themselves" afterward by attending Sunday afternoon orchestra concerts (which were "always attended by large audiences"). In a similar vein, Frankl (1972:64–65) tells us that "a kind of cabaret was improvised from time to time" in which there were "songs, poems, jokes" (some with satire), and that "The gatherings were so effective that [even] a few ordinary prisoners went to see the cabaret in spite of their fatigue even though they missed their daily portion of food by going." Furthermore, Cohen (1973:73–74) describes how in Westerbork people would go out in the evenings to a cabaret and listen to an orchestra, pay one another visits, and make dates, just "as in ordinary life." Despite their imprisonment and fear that they would be transferred to harsher camps, such practices enabled people to cope with the situation, as indicated by the following quotation: "But it just isn't true that you let it get you down for long. At least, not as far as I was concerned. For afterward [after the departure of the weekly train taking prisoners to worse camps] life went on as usual again."

Other accounts also discuss the significance and salutary effects of engaging in different kinds of ritualized behaviors. For example, Lengyel (1947:123) states: "No spectacle was more comforting than that provided by the women when they undertook to cleanse themselves thoroughly in the evening." Such a practice was important not only because it was a way of resisting the dirt and parasites, such as lice, but because "that was our only way of waging war against... our jailers, and against every force that made us its victims."

In a similar vein, Levi (1986:41) states that, while it was practically pointless to wash every day because the water was filthy, "it is most important as a symptom of remaining vitality, and necessary as an instrument of moral survival." Paraphrasing the advice of one of his fellow prisoners: "we must certainly wash our faces without soap in dirty water and dry ourselves on our jackets. We must polish our shoes, not because the regulation states it, but for dignity and propriety." Discussing the importance of a different kind of activity, Levi makes a similar observation about a rabbi imprisoned with him: "this thin, fragile and soft figure has managed to work for two years without falling ill and without dying, but on the contrary is lit up by an amazing *vitality* in actions and words and *spends long evenings discussing* Talmudic questions incomprehensibly in Yiddish and Hebrew with Mendi, who is a modernist rabbi" (p. 68; italics ours). Later, Levi also highlights the importance of another type of ritualized practice when he describes a group of tenacious, strong-willed, united Greek Jews who had managed to survive for three years in the camps and emphasizes how they "stand closely in a circle shoulder to shoulder, and sing one of their interminable chants.... They ... sing and beat their feet in time and grow drunk on songs" (pp. 71–72).

So too, in the account authored by Hénocque the value of RSPs such as regular discussions or engaging in religious confessions is clearly emphasized. Here, a priest states:

> Isolated from everything, abandoned by mankind, these men felt the need to rely on a superior power. But how many precautions were to be taken in order not to attract attention and therefore the wrath of the enemy. Sometimes we would talk in a block corner, sometimes we would discuss while walking—some other times, seated next to each other, my penitent and I, we would appear to be reading from the same book or looking at the images, and the poor captive, eager to receive comfort and pardon, would admit his faults, without any external sign of a confession. (1947:93)

Perl (1984:58–60) also provides us with a quite telling discussion of how she at one point in her confinement decided to initiate a plan of action that would take her out of her apathy and enable her "to remain a human being to the last minute of my life—whenever that would come." After retiring to their bunks in the evening, she began talking to her fellow prisoners about a variety of topics. Soon they responded and every evening they would hold conversations, recite poetry, invent games, tell stories, sing songs, and play a game she devised called "I am a lady . . . ," where they would imagine engaging in the various activities common to their prior ordinary lives such as shopping, lunching, and walking to the theatre and having supper afterward. She concludes by stating: "These evenings acted like a stimulant" providing them with a "human dignity" and "the moral strength to defend it with" (see also pp. 92–93 for additional discussion of how evenings spent with a group of friends talking about literature and singing prevented these individuals from breaking down).

The importance of ritualized practices such as diary writing and repeated discussions is also highlighted by Kammerer (1995:91), who provides this reflection on his internment: "Looking back, I realize—going over this little diary written with a pencil on bad paper—how much Block 26 allowed for an uninterrupted intellectual life, a real exchange of our pastoral and theological thoughts. From this standpoint, we were strongly resisting the general dehumanization desired within the camps, even if the material conditions were not ideal for this permanent 'recycling' [readjustment]."

Finally, we end this discussion with a quote from a secondary source that we have not included in our primary collection of accounts describing people's behaviors and experiences, since this work is an intellectual essay. Be that as it may, Amery (1980) provides a quite relevant observation about the activities and thinking of those who he was interned with. Although an agnostic, he confesses great admiration for his religiously and politically committed comrades and notes that their beliefs were a great help to them:

> Whether they were militant Marxists, sectarian Jehovah's Witnesses, or practicing Catholics, whether they were highly educated national economists and theologians or less versed workers and peasants, their belief and their ideology gave them that firm foothold in the world from which they spiritually unhinged the SS state. Under conditions that defy the imagination they conducted Mass, and as Orthodox Jews they fasted on the Day of Atonement although they actually lived the entire year in a condition of raging hunger. They held Marxist discussions on the future of Europe.... They survived better or died with more dignity than their irreligious or unpolitical intellectual comrades,

> who often were infinitely better educated and more practiced in exact thinking. (pp. 12–13)

As Amery's comments demonstrate, those who were best able to cope with and therefore adjust to the situation were the ones who persistently engaged in RSPs. By repeatedly participating in these quite different types of ritualized behaviors they derived a greater sense of stability, direction, and coherence in their cognitions and actions than did those around them (for a similar, albeit less detailed comment, see Lengyel 1947:110).

Cross-Cultural Research

The findings just presented concerning Nazi camp internees are totally consistent with what the first cross-cultural study of concentration camp internment accounts reveals (Knottnerus 2002). As previously mentioned, this study examined the accounts of prisoners in Japanese, American, and Soviet camps in addition to Nazi concentration camps. Here too one finds evidence of how the disruptive experience of internment led to deritualization and in some cases reritualization. Although far fewer accounts were examined for each culture, what is striking is how similar the experiences of quite different kinds of prisoners in very different cultural settings are.

While we will not discuss most of the findings of this study, a few narratives will be presented that demonstrate how consistent these accounts are with the findings generated in our more in-depth study of Nazi prisoners. We will restrict our presentation to a small number of very reflective, vivid descriptions by survivors of their experiences in Japanese, Russian, and American camps.

The debilitating effect that camp imprisonment had on American prisoners of war interned in Japanese camps is clearly conveyed in interviews with these prisoners. As in the Nazi camps we find evidence concerning the crippling effects of inactivity and loss of the will to live along with how ritualized activities, even the spreading of rumors, could be so important for a person's survival. Note the following comments from two soldiers about this very issue (La Forte, Marcello, and Himmel 1994:208): "I lived on rumors, and I'm sure the other fellows did, too"; and, "They just survived on them. I witnessed men actually giving up and dying. You could tell in conversations with a man that he had lost interest. Sometimes they would make remarks such as 'I just don't care anymore' ... and it usually wasn't long from the time we noted that they were giving up until the time they died. When I say not long, I mean two or three weeks usually until they were dead." This is the same behavior often noted in Nazi camps when writers refer to *Muselmann,* which is a term for those who have lost hope and will presumably die.

The value of ritualized activities, including more organized practices, is similarly emphasized by an American whose family was imprisoned in a Japanese internment camp. Booker describes some of these activities and how failure to participate in them had harmful consequences:

> Hardly a day passed after we took up residence in that warehouse prison before there arose a general demand for a program of camp activities: sports and games ... classes and lectures which would enable us to benefit from months and years that would otherwise be just dreary loss of time. Men who lacked the spirit to enter into the life of the camp just fell flat—like the one who sat through the days, his face in his hands, occasionally murmuring, "Oh, my God! Oh, my God!" ... That place looked hopeless; but we called it "Happy Garden," and set to work to convert it into a softball and football field, with an area for the amateur gardeners.... Happy Garden and its sports were a boon. During the months from March to September, camp activities centered there. That playground, and the University [classes prepared and taken by internees], and the occasional concerts and shows we put on, provided relief from camp tedium and kept up our spirits. (1945:183–84)

So too, prisoners in Soviet labor camps provide persuasive evidence about the value of simple ritualized practices, such as cleaning and exercising, and the harm to those who did not engage in such rituals. Witness the commentary of Bardach:

> Despite my fatigue and the cold, I kept the exercise routine I had followed at home and in the Red Army, washing my face and hands at the hand pump. I wanted to retain as much pride in myself as I could, separate myself from the many prisoners I had seen give up day by day. They'd stop caring first about their hygiene or appearance, then about their fellow prisoners, and finally about their own lives. If I had control over nothing else, I had control over this *ritual,* which I believed would keep me from degradation and certain death. (Bardach and Gleeson 1998:130; italics ours)

And from another Soviet concentration camp prisoner, Herling, we have the insightful observation about the value of what might seem to be quite mundane activities and the profound effect that not taking part in such rituals could have on people:

> All these activities had one quality in common—they imitated the normal occupations of a free life. Our behavior was a parody of the gestures, habits and responses of our former existence, observing the *symbolic ritual* of a dimly remembered routine.... The foregoing description, however, applies only to those few prisoners who made some effort to save themselves from complete demoralization. But the majority ... left their bunks during the evening only to fill up the hollowness of their stomachs with a pint or two of the inevitable hot water.... The majority of prisoners ... deluded themselves that this suicidal form of relaxation strengthened the organism of their bodies. In the camp the normal process was reversed: inertia and apathy hastened death, while any form of activity postponed it for an unforeseen period. (1951:143–44; italics ours)

Finally, we find comparable accounts about the importance of RSPs from Japanese Americans interned in American camps during World War II. Kitagawa (1967:71), for example, states: "The Pinedale Assembly Center thus took on a group personality and became a remarkably integrated community. There was ... no sign of moral disintegration to be seen all the time we were there.... The days went by fast because people were engaged in one program or another.... Church services, Scripture classes, and other religious functions, both Christian and Buddhist, were well attended." At another camp, however, the same writer describes a situation where women were extremely active but the men for various reasons barely participated in any ritualized activities:

> There were English classes, classes in flower arrangement, sewing, and Japanese music, and all sorts of cultural things going on.... they took active part in the PTA ... Issei women thus unwittingly became the happiest people in the relocation center. They even began to look younger ... the effect of life in the relocation center upon the Issei men was almost completely opposite to that upon the women. The men looked as if they had suddenly aged ten years. They lost the capacity to plan for their own futures, let alone those of their sons and daughters. It was indeed pathetic to see such moral and psychic (if not spiritual) deterioration develop. (pp. 90–92)

In sum, data collected concerning all of the camps reveal a wide variety of ritual types or contents and personal rituals, informal social rituals, and more formal organized ritual practices. And in all the cases studied compelling narratives confirm the beneficial effects of these rituals for internees. Indeed, for some, such practices made the difference between life and death. The wide variety of groups examined suggests the universal nature of these ritual dynamics.

Ritual Dynamics of Camp Life: Summary and Analysis

In essence, our investigation of Nazi concentration camps and the previous comparative research on camp internment show that disruptions are real and extremely powerful events in the lives of people, with those examined here being particularly severe and of a coercive nature. Various accounts demonstrate that such disruptions can result in extreme degrees of deritualization, negatively impacting individuals and groups in a multitude of ways. When this occurs the reconstitution of ritualized practices, that is, reritualization, buffers humans from and enables people to cope with this situation as reflected in numerous references to ritualized behaviors in all the accounts written by survivors. Furthermore, a significant number of narratives contain reflective statements about the importance of such practices.

These processes appear to be very basic to and widespread in the social lives of quite different groups, that is, among both French and non-French internees and various cultures as shown in the first study (involving both prisoners from the society sponsoring the camp and prisoners not from the society supporting the imprisonment facility). In other words, similar evidence concerning disruption, deritualization, reritualization, and self-reflective statements providing further corroboration for these occurrences are found cross-culturally—in an in-depth study of camps in one culture (Nazi Germany), among both French natives and those who are not French, among men and women, and among persons of different backgrounds and occupations.

What all of this research also shows is that although certain kinds of rituals such as religious practices are referred to by various prisoners during reritualization, wide variations exist in the content of practices discussed by internees. This is the case both between and among different cultural groups. Variations also exist within ritual types (e.g., different kinds of sporting practices, games, or religious activities). And these practices involve individualized rituals and overtly social practices of both an informal and a more organized nature. Furthermore, idiosyncratic practices are also described in certain accounts. Such practices attest to the diversity of ways in which ritual enactments may occur and how individuals can utilize very simple unique ritualized behaviors in ways that are significant to them and that provide a sense of stability and security to their lives. These practices underscore the meaningful, expressive, and consequential nature of rituals.

We would also point out that growing evidence concerning the wide variation in ritual types indicates that what is of overriding significance is the engagement in RSPs of any type, rather than what form these activities take. In other words, the content of practices very clearly appears to be of secondary importance to the need to perform ritualized behaviors.

Finally, a close reading and analysis of accounts by internees in concentration camps suggest there are different ways that the RSPs engaged in by them are meaningful and significant and create a sense of stability and security. While we do not consider this analysis to be exhaustive, we can at this point identify and categorize nine specific ways in which ritualized practices have a substantial impact on people. A close perusal of the quotes presented in this chapter and all the accounts analyzed in this research substantiates this assessment. In saying this we are not suggesting that every ritual practice is meaningful and important in all these ways. Practices vary enormously in their nature along with the situations they are enacted in, all of which influences how they are consequential for actors. Briefly stated, and not in any order of importance, they are as follows:

First, RSPs enhance social communication and relations. They can provide a conduit for socializing and interacting, and in doing so promote social cohesion among individuals and in groups. Second, ritual practices provide guidelines for, and thereby help structure or pattern, personal actions and group dynamics. This creates an order and stability in human behavior and social processes. Third, RSPs can create a sense of familiarity and closeness to previous experiences and events, including both social and personal experiences. By re-creating previously engaged-in ritualized practices actors regain to varying degrees a sense of grounding in their previous lives.

Fourth, ritualized behaviors can create and reinforce religious beliefs, faith, commitment, and feelings of fulfillment (as has long been noted by certain social analysts). These practices often involve a collective or social dimension of the religious experience. Fifth, RSPs may facilitate a sense of moral worth and dignity. By engaging in such practices people may achieve a feeling of pride or honor. Although this moral state may be part of a religious experience it need not be; it can be achieved through many different kinds of ritualized actions and accomplishments that are not religious in nature.

Sixth, ritualized activities can create a sense of discipline and control. By literally putting into practice the ritualized behavior, sometimes in opposition to another group or enemy, actors acquire a sense of empowerment. The individual knows that he or she can be in charge of and manage his or her actions, thoughts, and feelings. Seventh, the enactment of an RSP can produce a sense of vitality, enthusiasm, and dynamism. Such an energizing or life-affirming experience can be highly motivating, contributing to the perception on the actors' part that they are living up to their full potential and are pushing themselves to their limits. Eighth, engaging in an RSP can be mentally or intellectually stimulating. The enactment of certain kinds of ritualized behaviors can have a very salutary effect on people because they keep their minds active and alert, requiring concentrated thought and reflection. Ninth, ritual practices can create feelings of joy and elation. Oftentimes such excitement and delight is derived from social rituals that are conducted in an intense manner with others; they involve collective joy (for an interesting discussion of episodes in history involving collective joy, see Ehrenreich 2006).

Some of the RSPs we have identified that have these consequences are more overtly social while others are more personal, capable of being carried out in a more individualized or private manner. We would note in this regard that even in the case of the more individualized ritual forms, though, a social dimension is usually still present, albeit of a less explicit nature. Even when the act is conducted alone it may well involve meanings derived from social groups and/or be carried out in relation or opposition to other social groups and situations either in one's memory or imagination.

Again, this listing of the ways RSPs can significantly affect people in concentration camps represents a first effort to identify and categorize rituals in this manner. As such it provides the basis for further exploration of this topic. Future research into different types of disruptions, deritualization, and reritualization will try to more formally and comprehensively identify how RSPs are consequential for people not only in the types of camps examined here but in other very different kinds of social settings.

Indeed, these studies provide a framework that serves as a foundation for other research questions and endeavors dealing with how the breakdown of rituals affects people and society as a whole. For instance, one could hypothesize that as concentration camps, or any social environment, become less coercive and restrictive, the greater the number and diversity of rituals engaged in during reritualization. Diversity refers here to both the content of rituals and the individual, informal, and more organized ritual practices. Based on the assumption that rituals are vital to the adaptability and stability of actors, as conditions become less coercive we would expect that people become increasingly likely to engage in more and different kinds of rituals. They could more easily innovate and create new rituals or reconstitute previously engaged-in rituals. And it should be easier for individuals to acquire the resources needed to engage in different kinds of ritualized behaviors.

Or, building on the arguments of SRT, one could hypothesize that the higher the rank of RSPs enacted during reritualization, the more they enable actors to cope with a disruptive situation. If rituals have a greater impact on people's cognitions, actions, and social relationships as the rank of those practices increases (measured in terms of salience, repetitiveness, homologousness, and resources), we would expect higher ranked RSPs to have a more powerful impact on individuals who have experienced deritualization. They should be of greater significance for people and contribute to a greater degree to the stability of individuals and the groups they are in. This too is an issue that has not been studied.

Finally, research on concentration camps sets the stage for studies of disruption, deritualization, and reritualization in other kinds of social situations. Possible topics for study include the ritual dynamics involved in natural (noncoercive) disruptions such as disasters, macro level disruptions involving entire societies or regions of the globe, enforced (coercive) societal disruptions involving large numbers of people, disruptions in different cultures, other forms of disruptions such as long-term undertakings involving extreme isolation and stress, and disruptions of personal health due to medical problems. Research into some of these topics has in fact begun and will be highlighted in the last chapters.

In closing, as research continues along these lines, these are the types of issues that will be focused on with increasing specificity. Efforts such as these increasingly demonstrate the essential role rituals play in our daily experiences and how tenaciously humans seek out and create these kinds of behaviors that provide meaning and structure in their lives.

Chapter 6

Ritual as a Missing Link Within Sociology

Current and Future Research

All the research utilizing SRT presented in the previous chapters addresses the fundamental problem discussed in the beginning of the book—the failure to adequately recognize the importance of ritual and how it can be a vital tool for sociological analysis. Each of these studies of different groups is directly relevant to some, if not all, of the points enumerated in the first chapter concerning how rituals, and SRT in particular, provide a missing link in sociology. Indeed, this research and the various concepts and formulations discussed in the previous chapters, such as the levels scheme and the social conditions for ritual transmission, represent the first steps in a long-term effort to better explain how ritual dynamics play a crucial role in human behavior and society.

Let me identify in a more precise manner some of the ways in which work dealing with structural reproduction, strategic ritualization, and disruption and deritualization addresses the various points concerning ritual being a missing link in sociological thought.

SRT as a Missing Link

To begin with, these studies demonstrate that rituals do occur in both premodern and contemporary societies. The studies that have been either briefly described or discussed in a more lengthy fashion in the previous three chapters cover a wide period of time ranging from contemporary to premodern to much older social systems. Research has focused on the role of rituals in an ancient Greek society over two thousand years ago and more recently in nineteenth-century French society, the antebellum period in southern American history, and the late nineteenth and early twentieth centuries when the game of golf first emerged in the United States. And as we move closer to the present we have examined ritual dynamics operating in a number of countries in the early and middle parts of the twentieth century and more recently in Ireland and within various American ethnic groups. Finally, an experimental study of ritualization in small groups was conducted just a few years ago. In all of these groups, many of which are found in contemporary society, ritual plays a central role.

So too, we find that rituals occur not only in sacred or religious settings, as has been emphasized by scholars such as Durkheim and a good number of social anthropologists, but in secular contexts. In all of the studies discussed here, RSPs occur, at least to a significant degree, in secular settings. Whether involving educational institutions, recreational activities, warrior societies, urban communities, political systems, internment camps, and so on, the social phenomena are clearly nonreligious in nature. At the same time we do find religious elements in some of these cases and the unmistakable presence of religious rituals. Witness, for instance, how religion is intertwined with political and social motives in the parades of the Orange Order or how some prisoners coped with their internment in concentration camps by engaging in religious practices. But even in these cases the sacred dimension of ritual is part of or takes place in a setting that is quite secular in other respects.

We also find evidence in some of the groups studied of how rituals are inherently dynamic and capable of change. For instance, the studies on concentration camps show how rituals are sometimes modified and adapted by prisoners to the extreme, highly restrictive circumstances of these harsh environments. And, more generally, the different types of strategic ritualization identified by SRT demonstrate how agents can construct and tailor ritual practices to achieve diverse objectives within many different kinds of situations. This becomes particularly apparent when authoritarian political systems such as the Nazi party use their power to produce and alter rituals such as holidays, traditional commemorations, and celebrations to achieve their aims. And while no research was discussed dealing with the concept of transformative structural ritualization (discussed in chapter 2), this formulation explicitly argues that RSPs can be transformed and such change can occur in different ways.

Another way SRT provides a missing link for social analysis is in its examination of rituals occurring at different levels of society and how ritualized activities can be transmitted across those dimensions of the social order. The theory formally outlines a model of levels of social order and argues that rituals can spread both across and within levels. In that regard half of the studies focus on structural reproduction and show how RSPs can be transmitted from more macro to more micro levels of society, as in the case of rituals in educational organizations influencing the ritual dynamics of youth groups embedded in them.

A key part of this particular argument and related research involves SRT concentrating to a large degree on the everyday behavior of people, that is, the practices that constitute our normal, daily lives. By focusing on this dimension of social reality, SRT is especially relevant for a scientific understanding of the role ritual plays in our concrete lives as we move from one situation to another. Even rituals that occur at more macro levels of the social order may ultimately be implicated in and directly affect the daily activities of people. This point is underscored and supported by studies showing how ritualization processes can operate across and within different levels of the social order.

Moreover, SRT discusses how ritual can serve as a missing link to integrate or form a bridge between different traditions of sociological theory and research. Ideas from SRT can complement, be connected to, and/or enhance the explanatory power and arguments of various schools of thought. For instance, in terms of core assumptions and its overall perspective SRT is similar in a number of ways to and is informed by the web/part-whole approach developed by Bernard Phillips, Tom Scheff, and others, which calls for a clear definition of a problem, movement up language's ladder of abstraction, movement down the ladder of abstraction, integrating knowledge, and reflexive analysis (see Phillips 2001; 2007; 2009; Phillips and Johnston 2007; Knottnerus and Phillips 2009; Phillips and Christner forthcoming; Kincaid 1996; Phillips, Kincaid, and Scheff 2002; Scheff 1997; 2006).

In emphasizing a broad scientific approach to the study of social life both stress the significance of the concrete (or empirical evidence), such as the rituals people practice in social relationships, which SRT is concerned with. Both underscore the importance of developing abstract concepts, that is, moving up and down the conceptual ladder of abstraction between (a) concrete situations and (b) abstract theory. So too, both approaches give emphasis to the need for theory integration or synthesis along with the need to employ a variety of strategies in order to address the complexity of social behavior. SRT also recognizes the importance of reflexive analysis (i.e., researchers critically understanding how they are influenced by their social environment and cultural values), which is stressed by the web/part-whole approach. Finally, both stress the need to utilize knowledge generated by the scientific method to deal with social problems and inequities in modern society.

Other ways SRT has formed bridges to theoretical perspectives and traditions of research include using certain ideas from expectation states theory to better understand in the experimental investigation of structural reproduction how group leaders can play a crucial role in the reconstruction of group rituals. In addition, several of the studies described in this book expand our understanding of the ritual dynamics that can occur in ethnic communities, thereby contributing to work concerned with social processes and identity formation in such settings. And studies of ritual enforcement have begun to address how social power and ritual are often inescapably linked and consequential for each other. These and other efforts of an integrative nature will be focused on to a much greater degree in future research.

Another way SRT provides a missing link in sociological analysis involves its use of a wide variety of research methods. The studies that have been conducted employ both qualitative and quantitative methodologies, which have resulted in a wide variety of kinds of data. Methodological strategies range from experimental research to historical-comparative research and historical and contemporary case studies. Field research (observations), interviews, and an examination of archival documents are employed in numerous studies. Some investigations also draw on a close reading and analysis of secondary sources and in some cases the content analysis of primary sources. And several studies have drawn heavily on literary sources and accounts, including memoirs, biographies, autobiographies, novels, fictionalized autobiographies, diaries, and travelogues. Which of these types of accounts and literary documents are used depends of course on the particular subject matter and the kinds of evidence that are available for investigating the topic, for example, memoirs for the study of concentration camps or novels, biographies, and so on for research into nineteenth-century French schools.

These studies, therefore, employ multimethod research strategies, which attempt to benefit from the respective strengths of these different approaches. As opposed to relying on only one or a few research methods and kinds of data, these studies utilize a richer, more comprehensive array of evidence to corroborate the theory. Building upon this body of work, other research recently completed, currently underway, or planned for the future employs an even more diverse array of methodological strategies.

SRT also places great attention on how ritual influences social structure (or social organization), culture, and personality. The theory explicitly focuses on how ritualization drives much social behavior and results in the patterning of social relations, that is, the structure of relationships among actors. While this is apparent in all the studies, it is especially so in the investigations of structural reproduction where the ritualized actions, relations, and organization of the wider social environments lead to the creation of similar ritualized practices and structural arrangements in nested groups.

The same also applies to culture. In all the studies the role of rituals and the symbolic themes or meanings communicated through them, which provide direction to social life, are apparent. For instance, we see how important individually and collectively ritualized practices are for certain internees in concentration camps as they attempt to make their actions and relations with others more meaningful. So too, in other studies we see how power and ritual intersect as the latter helps provide focus and meaning in the lives of people within large-scale political groups such as the Orange Order and even authoritarian societies.

And we find that ritual influences personality structures. Although this work is in its early stages, findings from the concentration camp studies show how ritual can have consequences for people's ability to adapt to extremely disruptive experiences and through its effects on their cognitions, emotions, motivations, feelings of self-worth, and overall character result in a stable mental structure. For example, RSPs may be intellectually or cognitively stimulating. They may create a sense of control and self-discipline, thus empowering actors. Individuals may feel revitalized, energized, and more motivated by engaging in rituals. By reconstructing previously engaged-in, familiar rituals, individuals also acquire a sense of security. Furthermore, participation in ritualized practices can generate feelings of joy and happiness, that is, very strong emotional states of a positive nature. And ritualized behaviors can create a sense of dignity or moral worth, which can be a crucially important part of individuals' and groups' self-concepts or what are often thought of as their individual and collective identities.

Indeed, many of the effects of ritual are important for the development and maintenance of people's identity (and the collective identity of the groups they belong to), concepts sociologists and social psychologists are often concerned with. Here, however, I use the concept of personality to emphasize the idea that rituals can have profound consequences for individuals in a number of ways, impacting not only their self-concepts and social representations or identity but also their thoughts, feelings, motives, sense of security, and overall disposition or temperament.

Another way in which SRT can act as a missing link in sociology involves its use of a common analytical framework focused on the concept of ritual to explain individual and social dynamics involving behavior, social relations, and structure in different groups. All of the abstract concepts and formulations, which are logically consistent with and build upon one another, deal with ritual dynamics occurring in a wide array of empirical settings.

Let me in a few words catalogue most of the analytical concepts presented in this book. They include ritualized symbolic practices (and more basic concepts such as socially standardized practices, action repertoires, and schemas) and the four factors—salience, repetitiveness, homologousness, and resources—that determine the rank of RSPs. The concepts of embedded groups, domains of interaction, and structural isomorphism are defined along with formulations explaining how structural reproduction and transformative structural ritualization may occur. A model of ritual dynamics and levels of social order and the concepts of inter-level transmission and intra-level transmission are also presented along with a discussion of social conditions for ritual transmission.

Furthermore, the concept of strategic ritualization is discussed in addition to four specific types of strategic ritualization: ritual legitimators, ritual entrepreneurs, ritual sponsors, and ritual enforcers. Related to the latter, which is discussed in greatest depth, the idea of the program of RSPs is presented along with ritual program complexity and ritual enforcement complexity. It is suggested that ritual program and ritual enforcement complexity can vary with cases being identified at various points on a spectrum or gradation.

Lastly, concerning the interruption and breakdown of ritualized practices, the concepts of disruption, deritualization, and reritualization are presented. Related issues such as how RSPs involve the two key dimensions of action and meaning, the universality of ritual practices, and why the performance of rituals following deritualization are so important are elaborated on. Nine ways rituals can be meaningful and consequential for actors who have experienced the kind of coercive institutional disruption examined here are also identified.

Reflecting the points just raised, SRT provides a missing link because of its ability to address the complexity of human behavior and society. This approach can be employed to study many different dimensions of social reality in a wide array of investigations. In other words, a very wide scope of social groups and phenomena can be researched and analyzed using the ideas of this perspective. This is readily apparent in the studies described here, which focus on many different kinds of issues and groups ranging from the ritual dynamics of schools to the rearing of youth in an ancient militaristic society, the development of a recreational pursuit, status distinctions within slave societies, the ritual dynamics of different ethnic communities, parading by a political religious organization, techniques of domination in a totalitarian state, coercive disruptions and imprisonment, and the role of dance in collective events within an ethnic group.

SRT also acts as a missing link by demonstrating the importance of ritual in social life. Reflecting many of the points already made, this perspective reveals how rituals are not always reflective of other social forces and dynamics in society but rather are significant in and of themselves. Rituals are an autonomous dimension of human behavior that drives much of social life. This does not mean, however, that SRT denies the importance of other social forces and conditions. Recognizing the complexity of human behavior and society, other dimensions of social reality such as power and political systems, economic arrangements, status expectations, exchange processes, and occupational structures certainly exist and can both impact and be impacted by ritual. This assumption underlies some studies, which, as previously noted, integrate or form bridges with other perspectives emphasizing different aspects of social behavior. In fact, future theory and research in SRT will increasingly be of an integrative nature.

Finally, SRT demonstrates that rituals have great explanatory value. By recognizing the independent existence of rituals and the influence they can have on all of us we increase our ability to explain human behavior and social developments. As stated in the beginning of the book, by bringing ritual dynamics more to the surface in our perceptions and explanations of social life we are provided with another conceptual lens through which to view ourselves. In doing so, SRT contributes to the scientific analysis of society. Beyond that, however, through the insights gained from this knowledge we increase our ability to better understand our own personal and social experiences and improve the quality of our collective and individual lives (as will be discussed in the last chapter).

In sum, in a number of ways the concept of ritual and SRT provide a missing link in sociology. This perspective addresses and explores a variety of topics and concerns that much sociological research and theory (along with other social sciences) either ignore or fail to emphasize to a sufficient degree. That is why the work that has been described here creates the basis for further research and conceptual development in SRT as it attempts to broaden our understanding of how ritual dynamics play such a crucial role in human behavior.

Extensions and Future Research

Building on the studies described here more recent investigations and several new extensions or lines of research in SRT have begun to examine further issues involving ritual dynamics. I will highlight several of these developments, indicating in the process how this work points toward future research. This brief discussion will, therefore, focus on recent studies and likely research activities. It will be forward looking.

To begin with, research will continue in the three major areas dealing with structural reproduction, strategic ritualization and power, and disruption/deritualization. In regard to the reproduction of RSPs and social structure among groups, attention has recently turned to the issue of how consumer rituals can be transmitted to other sectors of society (Thornburg and Knottnerus 2008). Given the dominant role consumerism plays in modern societies such as the United States, consumer behaviors can be conceptualized as "consumer ritualized symbolic practices." Such ritualized practices, it is argued, can spread from the economic sector to other parts of society, shaping the orientations of actors and the nature of the groups they are in. While this can occur in different milieus such as education or sports another likely setting involves modern religious organizations. Recent research is exploring this issue and asks whether the phenomenon known as megachurches is a prime example of such a development, how such ritual dynamics operate, and what the specific effects consumer RSPs have on these religious organizations. The spread of consumer rituals from the economic sector to the religious sector and organizations within it involves the intra-level transmission of rituals throughout society and the inter-level transmission of rituals as they are reproduced within specific organizations and ultimately the subunits, networks, groups, and interaction among individual parishioners within these churches.

Other research focuses on strategic ritualization and power. Building upon studies that expand on issues discussed in chapter 4 such as how the organizational structure of the Orange Order has enabled it to strategically engage in parading and the consequences of such practices (Edwards and Knottnerus 2007; 2010), quite recent investigations have begun to examine two new topics. In one study we (Edwards and Knottnerus 2009) use two theoretical approaches—elementary relations theory (ERT) by David Willer and SRT—in an integrative manner to better understand how the Notting Hill Carnival in London has been strategically used in quite different ways by various groups to accomplish their goals. We argue that the festival is a collective ritual event that has played a crucial role in three different social arrangements discussed by ERT—coercion, exchange, and conflict—since its beginnings in London (and before that in Trinidad). This study directs attention to critical ways in which power, culture, and rituals intersect through the annual carnival.

In the second study (Berry and Knottnerus 2009), we use SRT to examine how the Indonesian state has used its political power to disrupt (or obliterate) the ritualized practices and culture of the indigenous inhabitants of Papua New Guinea. Through the strategic introduction of RSPs peculiar to Indonesian society and other kinds of disruptive interventions, Papuan society has in recent decades become increasingly threatened, leading to a number of strains and distressing consequences.

One further topic has also been examined using these concepts. In studying the community protest that emerged in Philadelphia's Chinatown in 2000 against a proposed sports stadium, Guan and Knottnerus (2006) examine the ritual dynamics that enabled this group to engage in such a movement and to achieve its objectives. We give special attention to how the biritual ethnic character of a more recent generation of Chinese Americans enabled them to strategically use RSPs associated with both Chinese and Americans to resist this development that threatened Chinatown. We also employ the theory to show how the protest movement used several dominant (highly ranked) ritualized practices, which contributed to its success and effectiveness in the community. This study should provide a foundation for subsequent research focused on the strategic use of rituals in social movements and how rituals can influence both the development and impact of collective episodes such as these.

Research dealing with disruption, deritualization, and reritualization is also expanding into the investigation of these kinds of developments in other kinds of groups and social settings. Studies by Thornburg, Knottnerus, and Webb (2007; 2008) have examined how people's everyday ritual practices are disrupted in disasters and the ways individuals respond to those situations. Based upon a content analysis of nineteen seminal sociological studies carried out by the National Academy of Sciences in the 1950s and 1960s evidence reveals many instances of deritualization manifested through its effects on action and meaning in rituals. Findings from this research also show that reconstituting RSPs after disasters enables actors to cope with such traumatic events. We identify the distinguishing features of these rituals, including the most common categories of rituals people engage in after disasters such as religious, visiting/comforting, discursive (verbal expressions such as singing and storytelling), family, recreational, and eating/drinking rituals. These studies strongly suggest that disasters not only result in structural and systemic disruptions of societies but also have profound social psychological and cultural consequences. By using the concept of ritual we are able to conceptualize and investigate the kinds of social dynamics that occur when disasters strike.

Recent research has also been investigating a very different issue involving the displacement of youth during China's Cultural Revolution (Wu and Knottnerus 2005; 2007; 2008). When educated youth during China's Cultural Revolution were forced in the 1960s and 1970s to leave their homes, which were usually in cities, and move to the rural parts of the country, such a move severely disrupted the ritualized practices that defined the daily lives of youth. By reconstituting RSPs individuals were able to alleviate the psychological tension and interaction problems caused by such disruptions. Our research examines the composition and significance of these ritual activities for youth. For instance, findings from the first study reveal that individuals engaged in both individual and collective RSPs including, respectively, (a) diary writing, letter writing, reading and studying, caring for pets, practices related to nature (such as walking in the forest), and personal activities such as exercising or playing music; and (b) storytelling, political rituals such as collectively reciting the words of Chairman Mao, singing and other collective musical events, raising pets, playing cards, and celebrating holidays.

In subsequent research we focus on the origins of the RSPs that people reengage in by examining in much greater depth one individualized practice: diary writing. Here we demonstrate how three factors play key roles in the formation of RSPs among these youth. An analytical framework composed of the wider social environment, the specific group context, and agents highlights how ritualized activities at the societal level provide the main source for the development of daily RSPs while, at the same time, characteristics of the local group and immediate situation both facilitate and limit the ways individuals interpret, judge, select, and modify what they are exposed to. Building on these studies, we (Wu and Knottnerus in progress) are currently investigating in much greater detail the origins, characteristics, and impact on youth of the collective ritualized practice of storytelling (this study will be in English unlike the previous research, which has been published in Chinese sociology journals).

Moving to a much more macro level of analysis, another study has recently focused on the importance of natural resources for ritual practices in East Asia (Sarabia and Knottnerus 2009). We employ SRT to analyze historical periods characterized by severe ecological degradation and disruption/deritualization. Such periods have often been referred to as "Dark Ages." The period examined corresponds to the former and latter Han Dynasties, a period that witnessed the writing of the first universal history of China and when discussion of ritual appears significantly in historical documents. Primary texts and secondary sources are used to investigate deritualization and the continuity, abandonment, and development of new ritual activities during this time period. Findings from our study show that despite periods of resource unavailability, which can last several centuries, social rituals survive. More specifically, by identifying how Chinese ceramics and burial practices were introduced into the Korean peninsula and Japan at this time, we see the strategies communities utilize when they no longer can obtain needed materials to engage in specific rituals. When faced with the loss of certain materials, societies such as these obtained alternative resources, which allowed them to create modified RSPs.

Finally, several additional projects are currently underway that also explore the ritual dynamics of disruption/deritualization (for a laboratory experiment studying disruption in group interaction, see Sell, Knottnerus, and Adcock, unpublished manuscript). For instance, Mason and Knottnerus (2009) are focusing on the impact of disruptive, long-term, stressful situations involving high levels of isolation on social behavior by examining the accounts of members of polar expeditions (and research stations). The goals of this research are to determine if RSPs are engaged in during expeditions, whether similarities or differences in ritual practices are observed, the types and frequency of these RSPs, the conditions that facilitate individuals' ability to engage in certain kinds of rituals, and the consequences of these practices for individuals and the group dynamics of the teams.

Although the number of accounts examined of expeditions in the late nineteenth and early twentieth centuries to regions such as the Arctic is fairly limited at this point, research clearly reveals that RSPs are present to varying degrees on all of the expeditions studied. Many similarities in ritual types have been observed across most of the sources with a few, such as celebrations of holidays and special events involving food/eating, being universal. Evidence also indicates that the enactment of rituals by the crews contributes to the social psychological success of the expeditions in terms of how they interact with one another and the morale among group members (social psychological indicators concerning the impact of rituals on group dynamics of the crew include social harmony, cohesion, cooperation, and stability). As additional data are collected we will continue to explore issues such as these and questions concerning how crew homogeneity and possible other factors may affect the likelihood that actors will engage in ritualized practices in these kinds of circumstances.

Finally, White and Knottnerus (in progress) are utilizing the lens of SRT to study the disruptive effects of learning that one is HIV positive. Viewing such a diagnosis as a type of disruptive experience, we are exploring whether those individuals who are best able to adjust to such a troubling experience, which presumably involves a high degree of deritualization, engage in a greater number of highly ranked ritualized practices. We are also collecting information through a survey about how others closely related to persons diagnosed as HIV positive respond to this situation in terms of the ritual dynamics they may engage in. Such findings could have important implications for different kinds of serious health care conditions, their impact on people, and how individuals can best cope with such experiences.

All of these studies of disruption, deritualization, and reritualization show in very different ways and among quite different groups how rituals are meaningful and consequential for human beings.

In addition to this work, several new lines of research within SRT have begun to develop. Some of this research has been conducted quite recently while other research is in its early stages. I will, therefore, briefly highlight four of these extensions. One development involves the study of the enactment, that is, activation or mobilization, of ritualized practices in formal organizations and communities. The first study of formal organizations focuses on the Enron Corporation and argues that rituals are an important part of corporate culture and can involve deviance (Knottnerus, Ulsperger, Cummins, and Osteen 2006). Utilizing multiple media sources, this research identifies symbolic themes expressed through daily work rituals at the company's corporate headquarters. The themes involve risk, gratification, pride, and fantasy. We argue that RSPs expressing these themes significantly influenced Enron employees, shaping their cognitions, behavioral patterns, and relationships. In doing so they led to the normalization and reproduction of deviant behavior. In addition to this study, other research focusing on the mobilization of rituals in communities is currently underway examining the role organizations play in sponsoring rituals in Asian Indian ethnic communities in the United States (Dasari and Knottnerus in progress).

Another extension has begun to address the issue of how special collective ritual events operate and influence actors' emotional states and commitment to a group. I argue (Knottnerus forthcoming) that in collective events such as rallies, religious ceremonies, and holiday or festival celebrations the greater the emotional intensity experienced by individuals, the greater will be their commitment to and integration within a group. A theoretical model identifies several features of the social situation that influence the development of emotions and commitment in these events. Four factors play a crucial role in the model: focus of attention, interactional pace, interdependence, and resources (several of these factors also contain subcomponents). This formulation is influenced by, and differs in certain ways from, classical analysts (especially Durkheim) and contemporary researchers including those working in the fields of social psychology and the sociology of emotions. Recent research using this model is examining how such processes can operate in a partially different manner even in virtual settings, where participation in online video games—massive multiplayer online role-playing games—can generate strong emotional feelings and a commitment to the virtual community people play in (Simpson, Knottnerus, and Stern 2009).

Additional research has also begun to focus on the role ritual dynamics plays in the creation and maintenance of social inequality, distinction, and exclusion. For instance, Mitra and Knottnerus (2004) have examined gender inequality and royal women in northern Indian society from 300 AD to 1200 AD. By focusing on some of the most important traditional rituals that governed the lives of royal women in patriarchal India, we show how gender-based inequalities were reproduced among royal women (even though they could have potentially used the power of their position to alter these arrangements). In another study focused on the behaviors of women volunteers in NGOs concerned with women's rights in modern India (Mitra and Knottnerus 2008) interviews with volunteers demonstrate how RSPs related to the traditional concepts of caretaking, sacrifice, and the concept of natural sexual differences continue to be emphasized by these upper and middle class women. Such processes, we argue, contribute to the reproduction of traditional gender roles, especially the traditional concept of ideal Indian womanhood.

And in an investigation of gender inequality in Malawi employing in-depth interviews, field observations, focus groups, analysis of second ary ethnographic research, key informants, and historical/secondary research, Minton and Knottnerus (2008) suggest that gender inequalities are partly due to the RSPs that define the daily lives of women and men. Despite political, legal, and economic reforms, significant forms of gender inequality persist in this country. This, we argue, is due to ritualized social behaviors that create and maintain women's unequal access to major resources, particularly in regard to occupations, education, and nutritional and agricultural resources. By focusing on these issues, this study contributes to underexamined aspects of social inequality involving social interaction and micro-level social dynamics. It also has clear implications for how efforts aimed at mitigating social problems such as these must address those social and cultural factors that can profoundly contribute to the persistence of social inequality.

Lastly, research has begun to focus on how ritual can influence the construction and maintenance of identity. Although in its early development, this line of research rests on the basic premise that RSPs can contain important beliefs and symbolic elements that contribute to the self-representations developed by individuals and groups. The aforementioned research on ethnic identity and biritual character among Chinese Americans (Guan and Knottnerus 1999; 2002) and the perpetuation of gender inequality in Malawi, which in part involves the maintenance of traditional female identity, builds on this basic idea. So too, research on Native American powwows rests on the assumption that ritual practices such as these involving the body and physical activities can influence the development of individual and collective identity. Future research will continue to explore such issues, focusing on how a variety of practices ranging from everyday ritualized activities to collective events and, in some cases, coordinated bodily movements can impact identity formation.

All of these extensions examine different aspects of the complex nature of social behavior. Be that as it may the potential exists for much more research that would expand on the range of topics deserving of study. For instance, a cursory listing of potential subjects would include how ritual contributes to the legitimation of social behavior and group conditions; how ritualized practices can serve as a mechanism of group attachment and commitment (e.g., radical conversion in certain religious or political groups, terrorist/suicide groups, or cults); the ways ritual transformation occurs; the role of the body and physical activities in ritual and the ways such enactments may be involved in the development of collective joy, collective identities, or the personality; and the analysis of how ritual dynamics can play a mediating role between agency and structure. These and possibly other topics will be the object of future analysis and study.

In sum, these lines of research and certain issues under study or of potential interest may be organized in terms of their analytical focus. Disruption/deritualization studies concentrate on the basic assumption that rituals are crucial to social life while emphasizing the symbolic dimension of ritual practices. Work on emotions, identity, and legitimacy directs attention to other key components of ritual behaviors in addition to their symbolic content. The analysis of ritual enactments in organizations and communities, structural reproduction, group attachment and commitment, and structural transformation directs attention to the impact of rituals on social arrangements and group dynamics (and vice versa). Research dealing with power and strategic ritualization, social inequality, and the transmission of consumer rituals further broadens the analysis of rituals and examines the ways they may affect critical conditions and developments within society. And the analysis of agency and structure focuses on the way ritual can act as a conduit linking the two.

Extensions and Future Research as a Missing Link

All of the extensions of SRT discussed here and possible research topics continue to be grounded in the proposition that ritual can provide a missing link in sociology. Of course, they do this in different ways. While it is not necessary to discuss this in detail, I will highlight a few of the ways this continuing work does so.

For instance, these studies examine ritual dynamics in both premodern and modern societies, as seen in research focusing on ancient China, ancient and contemporary India, contemporary American society, and modern British society. And the rituals examined occur in both secular and sacred/religious contexts, for example, megachurches, powwows, expeditions, a community protest movement, and a secular carnival. Certain of these studies also underscore how rituals are dynamic and capable of change including research showing how rituals such as burial practices were modified in ancient China and surrounding countries during a period of ecological degradation. Moreover, a number of these projects either explicitly or implicitly reveal ritual processes operating at different levels of the social order. For example, research on deritu - alization examines the impact of disruptions on individuals and social interaction, communities, societies, and even regions of the globe.

Several of these investigations are also of an integrative nature, involving the synthesis of different theoretical approaches and concepts. The development of a model addressing how collective ritual events can generate an emotional intensity that contributes to group solidarity draws upon concepts in SRT and work in social psychology and sociology dealing with emotions. In a different vein, research on Papua and Indonesia focuses on the interconnections of ritual and power, in particular of the state, and how the latter may strategically use ritual practices in the attempt to radically change another society. And the study of the Notting Hill festival utilizes theoretical concepts in elementary relations theory dealing with exchange, coercion, and conflict in combination with SRT.

So too, these studies employ a wide range of methodological strategies and types of evidence. Methods used include, for example, archeological evidence, surveys, the reading and analysis of written accounts such as diaries and autobiographies, interviews, field observations, focus groups, informants, historical research, newspapers, the examination of formal documents produced by groups or organizations, and secondary sources including studies by other researchers. These investigations also show

how ritual has consequences for social structure, culture, and personality. All of the studies demonstrate the power of culture by focusing on the central role of rituals in group life and the symbolic meanings expressed through them that define the experiences of people and their social worlds. Certain of the studies, such as those examining issues related to gender in Malawi and ancient and contemporary India, also reveal how important ritualized practices can be in creating and maintaining structured social inequalities. And the investigations of disruption, deritualization, and reritualization in a growing number of social situations are shedding further light on how ritual can have many consequences for actors' personality structure and identity.

All of these studies draw on a common vocabulary and framework to analyze ritual dynamics in various kinds of groups and situations. In doing so they demonstrate the relevance of this approach for better understanding the multifaceted, complex nature of human behavior and society. They provide further evidence concerning the crucial importance of rituals in social life and the explanatory value of formulations that concentrate on them.

Finally, these studies also contribute to our understanding of the social conditions involved in ritual transmission that were discussed in chapter 2. While these points are not formally discussed in the context of missing links in sociology, they certainly overlap with and are quite relevant to them because they focus on, among other things, the ways rituals are transmitted within and across levels of the social order and groups. For instance, research on Enron advances our understanding of how rituals are enacted, mobilized, and spread within formal organizations such as corporations. Investigations of social movements and collective ritual events also direct attention to the role social occurrences and processes can play in spreading and reinforcing rituals within and among groups at different levels of society. And studies of strategic ritualization and ritual enforcement enhance our appreciation of how social power can affect the transmission of RSPs throughout society whether that involves state governments, city and community organizations sponsoring collective events such as festivals, or other forms of influence and control.

Research and Key Concepts in Sociology

Finally, I would emphasize that SRT has implications for other studies, research issues, and theoretical concerns in sociology. While some of the previous discussion dealing with issues such as integrative efforts involving SRT and different theoretical and research traditions directs attention to this topic, this is a point that warrants repeating and more explicit attention. How ritual analysis can apply to other work in sociology may be conceptualized in different ways. One way stresses how the study of different research issues can be enhanced by investigations focusing on ritual. Another way of addressing this concern emphasizes how different theoretical concepts exist in sociology that are of direct relevance for SRT and that the latter can contribute to.

For instance, in regard to research, studies of the ritual dynamics involved with structural reproduction direct attention to the social dynamics of settings such as schools and classrooms. Research within the sociology of education and related areas examining classroom dynamics and educational organizations could benefit from an analysis grounded in SRT that focuses in a systematic way on the ritual dynamics that oc cur among both students and staff and the consequences such practices have for youth and schools (for examples of studies that to some degree illuminate the role of ritual in schools, see Kapferer 1981; Gehrke 1979).

Research addressing ritual dynamics and reproductive processes is also relevant for new institutionalism's focus on mimesis and how organizations model the forms of other social organizations (DiMaggio and Powell 1983). In this regard, attention to ritual dynamics would directly contribute to our understanding of how the "institutional logics" (Friedland and Alford 1991) of organizations are enacted and mobilized. This is because SRT provides a much more explicit focus on the role of actors in institutions, how they engage in RSPs, and the conditions that lead to the development of distinctive mind-sets and behavioral styles in organizational settings. More generally, SRT can help us better understand how ritualized behaviors permeate formal organizations and contribute to the development and maintenance of distinct organizational cultures, whether they involve corporate, governmental, health care, social service, or other organizational entities.

In a different vein, SRT helps shed light on how and why disruptions affect people the way they do and the nature of their responses to these experiences. For instance, there are well-known arguments that major interruptions in people's routines are individually disruptive, whether these changes are positive (e.g., marriage, beginning a new job, birth of a baby) or negative (e.g., death of a close friend or relative, loss of a job, divorce). In this regard, the Holmes and Rahe (1967) scale lists different stress events and gives scores for them. And, of course, Durkheim in *Suicide* (1966:246) made a similar argument suggesting that financial or industrial crises increase suicide, not because they cause poverty, but because they are "disturbances of the collective order." But, part of the problem with arguments about interruptions is the precision of these arguments and the many unanswered questions dealing with this issue. What is it about change that is disruptive? Are the effects of positive and negative events equated, and if so in what way? Are disruptions due to human or nonhuman causes, or is change involving only small-scale interaction versus entire communities or societies the same or different? SRT provides a theoretical framework to examine questions such as these by focusing on the role of rituals in human behavior, how disruptions of them can affect us, and the role they can play in coping with such changes.

SRT also has potential relevance for studies of social inequality/stratification. Research such as the Malawi study (Minton and Knottnerus 2008) that focuses on the ritualized behaviors that perpetuate women's lack of access to important resources highlights the importance of social interaction and micro-level social dynamics in the construction of structured social inequality. This is especially relevant because social psychologists have increasingly argued that theories and research that concentrate on the role social interaction plays in the development and maintenance of social inequality are limited (Howard 1994; Hollander and Howard 2000; Schwalbe, Godwin, Holden, Schrock, Thompson, and Wolkomir 2000; Harris 2000). Unfortunately, much research fails to give adequate attention to how interaction and the larger social context are related, for example, how the former may play a crucial role in the creation and reproduction of the latter.

Indeed, the study of social inequality should involve both structural factors (and historical conditions) and the role of everyday patterned behaviors and social processes. By focusing on either micro or macro levels of analysis (which many studies in social stratification do) the multifaceted and complex problem of inequality cannot be fully addressed. SRT specifically directs attention to one key dimension of everyday interaction—ritual enactments—and as previously discussed welcomes integrative endeavors with, among other things, approaches focusing on structural and more macro-level dimensions of phenomena such as social inequality.

Finally, studies addressing strategic ritualization expand our knowledge of how social power and social control operate in society. By examining the links between ritual and decision-making, organizational structure, and political processes, we enhance our understanding of how cultural and social psychological factors enter into the creation and maintenance of social power and the ways people are affected by its use. In addressing such issues we gain a better appreciation for how and why social life is structured the way it is from micro to macro levels of analysis.

These concerns are especially important because, as previously noted, researchers in recent years have begun to address how culture, power, and political decision-making intersect. Nevertheless, much more analytical and empirical research is needed on this topic. This is particularly so in regard to ritual and both the ways it is influenced by powerful groups and individuals and the impact it has on these groups and members of society. The systematic and cumulative study of such issues employing the theoretical concepts of SRT and various methodological strategies would enable us to far better understand the complex interplay between cultural and social psychological processes and power.

Another way of conceptualizing how the study of ritual is relevant to other work in sociology focuses on how various abstract theoretical concepts relate to and might be enhanced by SRT. A number of key theoretical concepts exist in the discipline which through the years have strongly influenced (either explicitly or implicitly) many sociologists' analyses of and investigations of social phenomena. Actually various scholars have identified and discussed what they consider to be the most important analytical concepts in sociology. For instance, Nisbet (1966) suggests that five unit-ideas have profoundly influenced the sociological enterprise: community, authority, status, sacred, and alienation. More recently, Phillips (see previously cited publications by the author) has persuasively argued that a number of theoretical concepts are crucial to sociological analysis and research. Among the concepts he emphasizes and uses in his work are cultural values, self-image, worldviews, social stratification, and bureaucracy.

I have found several abstract concepts to be quite relevant to SRT. In certain cases these concepts are directly related to some of the just-discussed research issues. Key concepts include bureaucracy, social inequality or social stratification, social power, community, social structure, social interaction, and identity (or personality).

For example, several current projects deal with bureaucracy in contemporary society. One project to be briefly discussed in the next chapter focuses on RSPs of high rank that shape the bureaucratic culture of organizations and the consequences of these bureaucratic arrangements for the persons who reside in them. Other interests focus on how bureaucratic systems can involve rituals (or mechanisms) of control that can be used for various purposes including acceptance of and compliance with the state and other social institutions. The concept of community has also recently played an important role in work discussed earlier in this chapter that presents a model for how various social factors can affect the emotional intensity of actors in collective events and their commitment to and cohesion in communities and other groups (Knottnerus forthcoming).

Other studies dealing with, for instance, women in India and Malawi are directly informed by the core sociological concept of social inequality. And the aforementioned research on strategic ritualization is obviously concerned with social power and some of the social processes contributing to the exercise of power in society. Furthermore, at the heart of SRT is a very strong concern with the core analytical concept of social structure. This is seen in its basic assumption that RSPs involve standardized (patterned) behaviors. The patterning of social behaviors and relations is at the base of much of the social order that characterizes interaction in our everyday lives along with other social occurrences. This basic assumption underlies much of the research discussed in this book such as the studies of structural reproduction, ritual dynamics in organizations, and deritualization and reritualization.

In sum, the ideas of ritual and associated theoretical concepts have been linked to a number of other important abstract sociological concepts. Much of this work that has been or is being done strives to develop concepts related to ritual such as rank, salience, repetitiveness, homologousness, resources, deritualization, strategic ritualization, and so on, along with empirical investigations that are directly related to core analytical issues identified by abstract concepts such as structure, bureaucracy, or inequality.

As work progresses, increasing efforts will be made to even more explicitly integrate and systematically connect the idea of ritual to core concepts such as bureaucracy, interaction, community, inequality, power, and structure. Such an effort is grounded in a basic commitment to the cumulative ideal and the scientific method's need to be open to all relevant phenomena in its investigation of human behavior and social life.

With that said, however, there is one important extension of SRT that has not yet been discussed. This last line of investigation focuses on applications of ritual theory and the relevance of this research for both social policy and our personal lives. It is this topic we turn to in the last chapter.

Chapter 7

Social and Personal Life

Applications and Implications

As stated in chapter 1, SRT is committed to working along a three-fold path involving theory, empirical studies, and application. In the previous chapters, especially chapter 6, seven specific lines of research and theory development were identified that are currently being carried out. These extensions focus on structural reproduction, strategic ritualization, disruption/deritualization, mobilization of rituals in organizations, emotions and collective events, social inequality, and identity formation. Applied research addressing social problems and inequities in society is the eighth line of research in SRT.

This extension deals with applied research grounded in SRT and the personal and social implications of this research and theory. Studies concerned with the applied aspects of SRT are in their early stages, so I will end with some general comments about possible applications and implications for people's lives. While much of this discussion is suggestive in nature the studies that have been conducted and my additional comments here point the way for future development. Just as the case with the research discussed in the previous chapter, future efforts will build upon this work and address in greater depth these issues.

A concern with the application of SRT is grounded in my own values and belief that sociology should address fundamental problems in contemporary society informed by the findings and implications of our research. For instance, studies of disruption, deritualization, and reritualization show that rituals can change. This is a point emphasized in a very recent essay where I note:

> ... what this research suggests is that ritualized behaviors can be interrupted, break down, be reconstituted, and sometimes change. And, when this occurs these changes can involve the reenactment of ritual practices which may greatly differ in both their nature and their consequences. In other words these newly created practices can range from having deleterious personal or social effects to their facilitating human survival and personal/social development. In regard to the latter possibility, reconstituted rituals may have beneficial or restorative consequences. This observation leads to the following proposal that once principles are well established and investigated, the next step would be to use this knowledge to alleviate undesirable social conditions. Such applied or change oriented efforts, grounded in a theoretical understanding of social life, could involve the development of interventionist techniques and policy recommendations. (Knottnerus 2009:77)

Indeed, as earlier noted, SRT is concerned not only with structural reproduction but also with ritual change or "transformative structural ritualization." Ritual dynamics can lead to new behaviors and social arrangements. And these changes can occur in different ways, in different settings, and at different levels of the social order. One way this can occur is through the promotion of new RSPs in settings such as formal organizations. The possibility of creating ritual change in this manner is actually addressed in the first studies dealing with the application of SRT and the development of social policy.

Applied Research

These studies are concerned with the mobilization of certain kinds of ritualized practices in formal organizations, as was the case in the Enron study. They focus on organizational deviance and how bureaucratic structures facilitate certain kinds of ritualized behaviors in the daily lives of people who work in formal organizations that provide care for the elderly (Ulsperger and Knottnerus forthcoming; see also Ulsperger and Knottnerus 2007; 2008; 2009a; 2009b). We argue that bureaucratic work rituals are an important part of social life for nursing home employees but that these behaviors can also lead to unanticipated, negative consequences involving maltreatment and neglect.

Our research indicates that certain RSPs that express bureaucratic themes dealing with staff separation, rules, documentation, and efficiency influence the behavior of employees and contribute to the unintended maltreatment of residents involving, for instance, physical neglect, medical dereliction, personal negligence, and bodily harm in both for-profit and nonprofit nursing homes in the United States. In essence, bureaucratic culture promotes ritualized practices and attitudes, which place bureaucratic goals over the needs of residents. These ritualized activities shape the daily lives of staff and residents within the organization.

Based on these findings various policy recommendations and strategies are presented that address these conditions. Many of these involve promoting and supporting new RSPs that counter the negative impact of bureaucratic rituals in nursing homes. Recommended alternative ritualized behaviors include requiring higher-level staff such as administrators to interact on a daily basis with residents, thus increasing communication and care, and training all employees to help residents with various kinds of problems and not just specific needs related to their formal duties and specific occupation (e.g., nurse, nurse's assistant, therapist, social director, supervisors and administrators, medications specialist, social director, etc.). These are just a few of the policy and procedural recommendations put forth based on research examining the ritual dynamics of health care facilities such as these.

This research, I would suggest, is extremely relevant because what it reveals is the complexity of settings such as nursing homes and the ways bureaucracy, organizational dynamics, and social behavior can intersect resulting in undesirable consequences for people such as residents in these institutions. Applications require dealing with the complexity of social life, a condition that is not necessarily advanced, if not drastically limited, by sociological specialization. By employing SRT with its focus on ritual in everyday life in social environments such as these we gain a fuller and more accurate understanding of one type of social problem.

Not surprisingly many other types of social problems and conditions could also be examined using the concept of ritual. Possible issues would involve the spread of consumer rituals through different institutions and organizations such as religious groups and megachurches, the misuse of power and rituals including the manipulation of rituals by totalitarian and other kinds of states in the modern world, behavioral and identity changes in prison inmates, and ritual processes contributing to social inequities including gender, racial, or class inequalities. Another concern focuses on how rituals could be facilitated that help people cope with disruptions (and deritualization) such as disasters, severe medical conditions interrupting individuals' lives, and long-term, stressful endeavors characterized by extreme isolation such as expeditions or space missions.

For instance, research on disasters suggests that more attention should be given to assisting victims to reconstruct RSPs after disasters. Efforts aimed at helping people engage in reritualization, thereby reestablishing some sense of normalcy and stability in their daily lives, could greatly facilitate the recovery process. It is also possible that rituals are very important for emergency workers, helping them to manage and cope with the pain and distress of their work. Simple examples of the latter might include creating regularized break periods where workers can interact and talk, holding periodic social events and celebrations that highlight accomplishments and progress being made, or small ceremonies such as stopping work and standing in silence when bodies are removed from the location (as actually occurred among rescue workers at the 9/11 site).

Studies of the spread of consumer rituals to religious organizations and how this may impact religious groups such as megachurches could also provide information about the potential negative consequences of such developments. This knowledge might then be used to counter or modify some of these processes. For example, ritualized practices expressing symbolic themes that are not consumer based, which may be grounded in the religious group's long-established beliefs, could be reinforced through sermons and special events, smaller group activities, and organized projects within the church.

Finally, research examining how the diagnosis of being HIV positive affects both victims and family or friends who are close to them should help us better understand the significance of rituals for adjusting to disruptions involving our personal health. Knowledge about the role reritualization plays in the personal and social lives of people experiencing serious medical and health problems could enable caregivers and others to develop far more effective support services and forms of assistance.

Theory and studies in SRT will give increasing attention to issues such as these and the ways ritual practices can play a fundamental role in the creation of undesirable behaviors and conditions along with how an understanding of ritual dynamics can provide the basis for reducing, if not eliminating, them. While much of this work will continue to be addressed to academic audiences, particularly researchers and students, it will also be directed to other professionals and a more general readership in order to bring about more productive policy changes and successful interventions.

Personal Implications

I close this chapter by discussing the personal relevance of SRT, because this perspective has a number of implications for people's behavior. This discussion will be of a more general nature emphasizing how an aware ness of the role rituals play in human behavior and society can benefit all of us and potentially enhance the quality of our lives. These suggestions rest upon the arguments and findings of SRT concerning the need for ritual in our lives, its inevitability, how it creates meaning and a sense of stability, and the significance of the kinds of rituals a person engages in (as seen, for example, in the nine points concerning the consequences of rituals for concentration camp prisoners). I will highlight five points concerning the personal relevance of rituals.

To begin with, findings from research suggest the importance of being aware of the influence rituals derived from sources outside of the individual have over us. Rituals can be a part of the society or subculture we live in, shaping for better or for worse the ways we behave and think on a daily basis. While we may not be aware of their presence or consciously think about how they affect us, their impact on what we do, believe, like, or dislike can be immense. Groups or other individuals may also deliberately encourage rituals, which then unwittingly shape our actions. The transmission of rituals may or may not, therefore, be strategically planned and promoted. Either way the consequences are real!

The model of levels of the social order presented in chapter 2 can help us better understand this point. RSPs originating from distant dimensions and parts of society can be transmitted through different levels of the social order down to the personal level, ultimately shaping our interpersonal relations and individual experiences. Indeed, the reproduction of RSPs through different levels—from macro to micro—can ultimately influence the ritualized practices that define our taken-for-granted, everyday lives as we move from one situation to another. In this way they can shape the way we think and behave. So, the influences can come from afar and not be clearly seen, understood, or even known. Yet they may have a profound effect on who and what we are.

For instance, ritualized ways of thinking and acting may be promoted by business groups and advertisers for economic and marketing purposes or they may be encouraged by political groups in the attempt to influence people's support for or opposition to different policies. The sources of these RSPs may involve large-scale organizations operating at the most macro levels of society if not the globe. We may never personally come into contact with these groups and the individuals who run them, but the practices they promote are transmitted to each of us within society through different mechanisms such as the news media, movies, television, magazines, and the internet and through different levels of the social order including various kinds of organizations, community ties and associations, different social networks, and face-to-face interaction with others.

While extremely totalitarian societies are probably the clearest illustration of political groups and individuals using rituals, one case being the Nazi's enforcement of them, examples also exist in modern democratic and nondemocratic social systems. Notice, for instance, how often elected and nonelected political leaders will associate themselves with military, civic, or patriotic organizations as they participate in military events and national/civic commemorative occasions and celebrations (e.g., speaking before military groups, taking part in national events such as Fourth of July celebrations, attending religious gatherings that celebrate one's country, and participating in commemorations for famous leaders, past wars, and the war dead). In doing so they link themselves with and reinforce certain symbolic themes associated with the collective ritual event and the groups participating in it, such as patriotism, an emphasis on allegiance and obedience to the government, pride or hubris for a country, the perceived importance of a country's national/international power, and the moral or sacred worth of a society, its way of life, and its political/economic system. Usually these ritualized deeds and images are quickly transmitted through the media such as newspapers, television, magazines, and the internet along with more informal communication channels involving friends and associates or individuals throughout society, which in turn may reinforce their beliefs and support for these political groups and leaders.

Other ways this can occur involve the transmission of consumer rituals within American society and the world, which leads people to purchase, if not over-purchase, material goods and commodities such as the newest appliances, electronic devices, or computer games, a new car every several years, the most recent cell phone, or bigger and more expensive houses. Originating out of the economic/market sector and the businesses and corporations that play such a major role in contemporary societies such as the United States, consumer rituals can spread through other organizations, the workplace, recreational groups, clubs, churches, schools, and ultimately the array of social networks and informal groups surrounding each of us including our peers, friends, and families. These ritualized practices and beliefs, stressing the importance of consumer items and our need to acquire such goods if we are to feel gratified become part of our taken-for-granted lives. They shape our most basic assumptions about life, which become so second nature to us that many of us never even stop to question and reflect about their validity and consequences. It is such ritualized behaviors and orientations that can lead to the very real, personal financial problems we see today, for example, people of all backgrounds and persuasions falling into debt, being financially overdrawn, possessing little or no savings, becoming unable to meet their credit card and loan payments, facing the foreclosure of their homes, and experiencing the stress and hardship resulting from these developments.

Of course, research has also shown how ritualized practices can simply be a part of the society and groups we are members of and the powerful impact these rituals can have on all of us. While not strategically promoted, these rituals can be found in all kinds of social environments. These ritualized behaviors and arrangements can profoundly influence the ritualized practices that develop among individuals and groups embedded in these social environments. Examples range from youth groups and students being affected by the ritualized practices found in the educational institution they study and perhaps reside in to problem-solving groups, work teams, and research and development groups being influenced by the ritualized ways of thinking and behaving found within the larger corporate or bureaucratic governmental organization they work in. As research has also shown, sometimes this can involve different kinds of deviant ritualized behaviors that permeate organizations and shape the perceptions and actions of people who are in them. While we have found this to be the case in studies of a large corporation (Enron) and health care facilities (nursing homes), deviant rituals may well develop in many other settings such as governmental policy groups, university departments or committees, community groups, sports organizations and teams, and government agencies (for an insightful study of the taken-for-granted aspects of organizational life and the unwarranted acceptance of risk in the federal agency NASA, see Vaughan 1996). All of these examples attest to the importance of being aware of the impact that rituals derived from groups and society at large have on us.

Second, in a related vein, sometimes groups may actively encourage intensive exposure to ritualized practices among its members in the attempt to control their beliefs and actions. Such authoritarian and/or ideologically driven groups (or what are sometimes referred to as high-demand groups or totalistic groups), I suggest, often employ rituals in a very focused manner in the attempt to shape the cognitions and behaviors of its members. While this could involve entire societies such as ancient Sparta or totalitarian political systems, I raise this as a separate point to emphasize how there may be smaller groups in contemporary society who also strategically use rituals to influence people, usually group members.

This is a topic that has not yet been formally studied (but hopefully will be in the future). A cursory examination of such groups, however, suggests some of them use rituals in a deliberate way to influence and control people. Examples of such occurrences could range from dogmatic religious groups or political movements to various sects and cults. Of course great variations exist among extremist or high-demand groups in terms of their goals and motives. Note for instance the differences between a terrorist group such as Al Qaeda and totalistic (nonterrorist) groups such as the Moonies, the FLDS (Fundamentalist Church of Jesus Christ of Latter-Day Saints), or more mainstream religious groups who are extremely strict and demanding. Yet, ritual dynamics can be of great significance in all of these groups and help create and sustain a person's identification with and commitment to them. That is why an awareness of the role rituals play in groups of this kind are so important, both because we or others we know may belong to them and because we may be, for better or worse, affected by their actions.

A third point is based on studies dealing with disruption and deritualization that demonstrate how disruptions of our lives and the groups we are in such as families, friends, coworkers, and communities can have problematic consequences. This research clearly shows the importance of reconstituting old or new rituals when such disruptions occur in our personal lives. Indeed, all of us are subject to many kinds of interruptions in our lives—negative, positive, human in origin, nonhuman in origin, and so on. Such events can range from disasters, divorces, breakups, and the deaths of family members or friends, to changing one's residence, births, marriage, and major changes such as losing a job or acquiring a new one.

When such disruptions occur people often experience confusion, uncertainty, or bewilderment, and strong emotional responses such as sadness, depression, anger, resentment, frustration, or grief. Some lose a sense of direction, withdraw, and become inactive. Others engage in personally unproductive behaviors such as drinking, drug taking, gambling, or continually watching television or movies as a way of diverting their attention from the troubling disruption they have experienced. On the other hand, some inadvertently or based on "folk wisdom" begin to reengage in some kind of productive activity on a regular basis. SRT clearly emphasizes the importance of the latter and the need for people to reconstitute ritualized practices in their personal and social lives when they experience such disruptions. The reconstitution of rituals in our lives subsequent to such events can be extremely valuable because such behaviors, as previously discussed, can provide a sense of purpose, focus, and direction along with feelings of security, familiarity, and stability. Which ritual practices are significant to individuals undoubtedly vary, for example, work, religious activities, hobbies, meeting with friends, caring for a pet, play or sports, musical or artistic activities. But what is important is that we begin to reenact meaningful rituals on a regular basis when these situations occur.

Fourth, the dynamics of rituals are significant. Several studies grounded in SRT have shown that rituals are dynamic, in that they are subject to change. While the following has not yet been researched, it is also possible that some rituals over time become mere routine or habit, losing their import or value. They may become deadened, lacking in meaning and import. By being aware of this possibility an individual is better positioned to respond to such a development and cultivate different and more meaningful ritualized practices or to take steps to reactivate and reenergize the rituals they are engaging in by altering them in some way. As a simple example, if the daily ritual of exercising starts to lose some of its appeal, one might develop new exercises to perform, alter the time or order one does them, or change the location where he or she exercises. Or if one finds she or he is losing interest in reading every day the individual might energize this ritualized practice by choosing a very different type of reading material such as reading biographies, autobiographies, or short stories instead of novels. While major changes may be called for, sometimes small modifications of ritual practices can prevent such activities from losing their significance.

So too, under certain circumstances what is routine may be revitalized and become a meaningful ritual in the life of a person. We saw this in the concentration camp studies where what seemed to be trivial behaviors such as cleaning oneself became a ritual practice that held great meaning for some prisoners and enabled them to survive. It is possible that under certain conditions such a development could also occur in our normal lives. For instance, if personal economic loss suddenly prevented a person from engaging in certain expensive recreational practices (e.g., skiing, boating, collecting rare and expensive items), one might find that what had been previously considered minor or mundane activities could become important RSPs providing meaning and direction in his or her life (e.g., going for daily walks, growing a food or flower garden, bicycling).

Finally, it cannot be emphasized enough that the types of ritualized actions we engage in on a daily basis can lead to very different personal and social outcomes. Our ritualized behaviors may, for instance, result in meaningful interaction, closeness, and effective communication with others or social isolation, estrangement, and what some refer to as alienation. They may lead to personal growth or rigidity and an accompanying sense of fulfillment or deprivation and loss in one's life. So too, the ritual enactments that define our lives may result in a sense of autonomy and self-determination or a dependency on and manipulation by others, that is, a control or lack of control of one's life. Stated more generally, the personal and social rituals that shape so much of our behavior can result in either unproductive, unconstructive, harmful consequences or beneficial, constructive outcomes that can enrich and expand our own lives and the lives of others.

Only one other individual that I know of has addressed this issue. Bernard Phillips (see especially Phillips and Christner forthcoming) has emphasized how our worldviews and cultural values can profoundly impact our behaviors and the quality of our lives. He contrasts a bureaucratic worldview and bureaucratic rituals that involve limited interaction and communication, hierarchical relations, and specialization with an evolutionary (or interactive) worldview and rituals that are meaningful and expressive. It is the latter, he argues, that can create a sense of control and freedom in our lives. His discussion of these issues is couched within a more encompassing argument referred to as the east-west strategy that presents a strategy or technologies for personal and social development based on the use of a broad scientific method in our everyday lives.

Although SRT has not in any way addressed these issues in such a comprehensive and focused manner, affinities do exist between the ideas and findings of this perspective and the arguments of Bernard Phillips. And work within SRT does have implications for how to achieve certain outcomes through the rituals we engage in. To illustrate, findings from studies of concentration camps have identified a number of ways rituals were meaningful and consequential for prisoners. What we might refer to as positive or constructive RSPs would be rituals that in our everyday lives have some of these same benefits that they had for camp internees. Such rituals could, for example, enhance social communication and relationships, feelings of moral worth and dignity, a sense of discipline or control, a sense of vitality, intellectual stimulation, or feelings of happiness and possibly elation. And various techniques or procedures, derived in part from the arguments of SRT, could be developed for achieving some of these desired outcomes and more generally the outcomes discussed by Phillips.

For instance, just as groups may engage in strategic ritualization and employ or enforce programs of RSPs in society, individuals can also construct programs of rituals in their personal lives. In other words, we all possess the potential for developing personal programs of RSPs that can help us achieve desired outcomes. Of course, saying this is quite different from doing this. Various obstacles may stand in the way of putting into place such a program, beginning with a lack of understanding of what ritual is and does. For that reason various techniques or strategies may be called for that increase the likelihood that we can actually develop a personal ritual program. Certain of these techniques involve some of the previously discussed points dealing with the personal implications of SRT. But they are of such importance for the development of a ritual program they bear repeating within this context.

To begin with, it is incumbent on us that we develop an awareness of ritual's place in our lives and how it can lead to desirable or undesirable consequences. Without becoming sensitized to and reflecting about its occurrence, we remain oblivious to ritual's existence and impact on us. We also need to become aware of how social arrangements and groups, as previously discussed, can influence us. When, for example, we are in embedded groups we are susceptible to the influence of highly ranked RSPs in the wider environment. So too, we are always susceptible to groups in society that strategically promote RSPs in our lives. That is why it takes extra effort and measures to counter the influence of ritual dynamics such as these, beginning with the need for developing a critical awareness of these social forces. We also should be aware of the potential for disruptive experiences such as the deaths of loved ones, disasters, job loss, economic problems, and so on to interrupt the rituals we normally practice and how we can "lose our bearings" when such events occur. Because such disruptions can be so emotionally and mentally upsetting there is always the possibility we will not reestablish ritualized practices that are beneficial to us after such events. That is why it takes concentration and effort to effectively cope with developments such as these.

It is also important that we recognize that ritual practices can sometimes become routinized and habitual, thus losing their significance. Here too, effort and thought is needed to keep rituals meaningful and vital. Finally, we should develop an awareness of how unconstructive activities can "sneak" into our lives and become ritualized. Without our even knowing it we may begin to engage in various kinds of rituals that do not serve our better interests. For instance, in coping with stressful, troubling situations of different kinds such as being rejected, stigmatized, or mistreated by others we may develop new rituals that seem to provide some compensatory gratification but are harmful to us in the long run. A not uncommon example involves the person who begins to ritualistically overeat as a way of coping with some problem or disappointment one has encountered (I wish to thank Krista Schumacher for pointing out how we can develop coping rituals such as this).

More precisely, a *personal program of RSPs* involves the development of an array of RSPs engaged in on a daily basis in different situations that enable individuals to accomplish goals shaped by the values they possess. To do this requires an awareness and at least some understanding of the place ritual occupies in social behavior, how ritual dynamics operate, and the social situations and forces that can impede or facilitate the development of RSPs. Along with that it is important to understand what the cultural values one holds are that determine the goals sought by the ritualized practices one is fostering. Lastly, there is the need to strategically develop the means necessary for effectively implementing a program of RSPs.

While there are potentially many ways to increase a person's chances of success, let me highlight just a few basic points concerning how ritualized practices can be more successfully implemented. To begin with we can strategically choose to interact with individuals and groups and be in social situations that facilitate and reinforce these rituals, whether that involves the settings where we work, clubs, sports groups, friends, cliques, people we are in relationships with, or marriage partners. Because humans are social creatures, the social environments we are in can profoundly affect how successfully we engage in a program of RSPs.

At the same time, we can intentionally avoid social situations and groups that reinforce undesirable RSPs. One of the easiest ways to lose our focus and commitment is to interact with others who practice quite different rituals, for example, interacting with a peer group that devotes most of its time to socializing by drinking at bars or using Twitter when one's goal is to also engage in ritualized behaviors such as playing or enjoying music, discussing political and social issues face to face, practicing yoga, playing a sport, gardening, working out, or reading.

Attention to the resources needed to practice rituals is also useful. We can only perform rituals if we are able to acquire the resources that make this possible. This means we may have to learn how to engage in RSPs that our limited resources allow for. A simple illustration of this point would involve choosing to exercise at home with used equipment instead of going to an expensive health club beyond one's means.

Finally, we can support through word and action (e.g., expressing one's opinion, volunteer work, financial contributions, petitions, voting, and surveys) policies, political and social groups, political and civic leaders, and government legislation that promote constructive rituals throughout our communities and society as a whole. Examples here could range from supporting the development of environmentally friendly ritualized behaviors, energy-saving practices, public events and financial support promoting literacy and reading, recycling, and ritualized activities that improve public health, to collective events such as festivals, parades, and celebrations that enhance social communication, social tolerance, and the social health of the community or city we live in. Such strategic actions naturally go beyond the personal level but they still have consequences for each of us as individuals. By promoting constructive rituals we foster a social climate and structural arrangements that we all live in and will benefit from while at the same time reinforcing our personal commitment to desirable ritual practices through the very act of expressing our support for them.

These ideas will be expanded upon in subsequent work, but the points presented here, at the very least, indicate how we can begin to develop a fuller awareness of ritual's relevance for all of us. They also show how we can more productively employ rituals through personal programs to realize our values and goals, especially for achieving a richer, more humane personal life and social world.

Conclusion

In summary, SRT contributes to the explanation of ritual in everyday life. It is a growing program of theory and studies that seeks to expand our understanding of the ritual dynamics operating in various social situations and across different levels of the social order while forming bridges with other perspectives and research in sociology.

In numerous ways, as has been emphasized and discussed throughout this book, it provides a missing link in sociology. For this reason, the concept of ritual and SRT helps us to explain some of the complexity of human action and social life. While most of our attention has focused on how this perspective can be used to better understand how ritualized practices and structural arrangements are reproduced in groups, how rituals can be strategically used and influenced by social power, and how the breakdown and reconstitution of ritualized behaviors can be so consequential for human beings, other issues and developing lines of study have also been highlighted and described. All of this work suggests that SRT's scope continues to expand and should lead in the future to many more contributions that further demonstrate how it can provide an important link in sociological analysis and research.

At the same time, SRT is relevant to different research issues and theoretical concepts in sociology such as bureaucracy, social structure, interaction, inequality, power, and community. Future work will more formally address how this approach can be applied to various research problems and can be systematically linked with crucial theoretical concepts within the sociological literature.

Finally, the arguments and findings generated by this body of research have numerous implications for people's lives. This work provides the basis for policy recommendations dealing with various issues including the running of formal organizations and health care. By increasing our understanding of this dimension of human behavior we are also better able to appreciate, alter, and design ritualized practices that may either enhance or diminish the quality of our own personal experiences and that originate from or are influenced by ourselves, others, or events beyond our control.

A commitment to theory, empirical research, and applying this knowledge to social problems, social policy, and our personal lives will, therefore, continue to drive the efforts of those using this approach. This work should contribute to a growing body of knowledge about ritual and the ways it is an essential part of human behavior and society, past and present.

References

Alexander, Jeffrey C., Bernhard Giesen, and Jason L. Mast, eds. 2006. *Social Performance: Symbolic Action, Cultural Pragmatics, and Ritual.* Cambridge: Cambridge University Press.

Allen, Paula Gunn. 1986. *The Sacred Hoop: Recovering the Feminine in American Indian Tradition.* Boston: Beacon.

André, Robert. 1978. *L'enfant miroir.* Paris: Gallimard.

Audoux, Marguerite. 1911. *Marie-Claire,* translated by John N. Raphael. New York: Grosset & Dunlap.

Axtmann, Ann Marguerite. 1999. "Dance: Celebration and Resistance: Native American Indian Intertribal Powwow Performance." Ph.D. diss., New York University.

———. 2001. "Performative Power in Native America: Powwow Dancing." *Dance Research Journal* 33:7–22.

Baird, Jay W. 1990. *To Die for Germany: Heroes in the Nazi Pantheon.* Bloomington: Indiana University Press.

Balzac, Honoré de. 1948. *The Short Stories of Balzac.* New York: Dial Press.

Bandura, Albert. 1986. *Social Foundations of Thought and Action: A Social Cognitive Theory.* Englewood Cliffs, NJ: Prentice-Hall.

Bardach, Janusz, and Kathleen Gleeson. 1998. *Man Is Wolf to Man: Surviving the Gulag.* Berkeley: University of California Press.

Barrès, Maurice. 1897. *Les déracinés,* edited by Eugène Fasquelle. Paris: Bibliothèque-Charpentier.

Beck, Reinhart. 1997. "National Socialism." Pp. 622–27 in *The Encyclopedia of the Third Reich,* edited by C. Zentner and F. Bedurftig. New York: Da Capo Press.

Bedu, Jean-Jacques. 2001. *Francis Carco: Au cœur de la bohème.* Paris: Editions du Rocher.

Bell, Catherine. 1992. *Ritual Theory, Ritual Practice.* Oxford: Oxford University Press.

Bellah, Robert N. 2005. "Durkheim and Ritual." Pp. 183–210 in *The Cambridge Companion to Durkheim,* edited by Jeffrey C. Alexander and Philip Smith. Cambridge: Cambridge University Press.

Bennett, Billie Ruth. 1991. "The Role of the Native American Pow wow in the Acculturation Process." Ph.D. diss., University of New Mexico.

Berger, Joseph, M. Hamit Fisek, Robert Z. Norman, and Morris Zelditch. 1977. *Status Characteristics and Social Interaction.* New York: Elsevier.

Berger, Joseph, and Morris Zelditch, Jr., eds. 1993. *Theoretical Research Programs: Studies in the Growth of Theory.* Stanford, CA: Stanford University Press.

———. 1998. *Status, Power and Legitimacy: Strategies and Theories.* Brunswick, NJ: Transaction Publishers.

Berger, Peter L., and Thomas Luckmann. 1966. *The Social Construction of Reality.* Garden City, NY: Doubleday.

Bernardi, Fabrizio, Juan J. González, and Miguel Requena. 2007. "The Sociology of Social Structure." Pp. 162–70 in *21st Century Sociology: A Reference Handbook,* vol. 1, edited by Clifton D. Bryant and Dennis L. Peck. Thousand Oaks, CA: Sage.

Bernhardt, Sarah. 1968. *Memories of My Life: Being My Personal, Professional, and Social Recollections as Woman and Artist.* New York: Benjamin Blom.

Berry, Phyllis, and J. David Knottnerus. 2009. "Ritual, Collective Identity and the Modern State: Indonesia and the Reconstruction of Papuan Society." Presented at the annual meeting of the American Association of Behavioral and Social Sciences, Las Vegas.

Bessell, Richard. 1980. "The Rise of the NSDAP and the Myth of the Nazi Propaganda." *The Wiener Library Bulletin* 33:20–29.

Bienenstock, Elisa J., and Phillip Bonacich. 1993. "Game Theory Models for Exchange Networks: Experimental Results." *Sociological Perspectives* 36:117–35.

Blau, Peter M. 1981. "Diverse Views of Social Structure and Their Common Denominator." Pp. 1–23 in *Continuities in Structural Inquiry,* edited by Peter M. Blau and Robert K. Merton. Beverly Hills, CA: Sage.

Blumer, Herbert. 1969. *Symbolic Interactionism: Perspective and Method.* Englewood Cliffs, NJ: Prentice-Hall.

Booker, Edna Lee (in collaboration with John S. Potter). 1945. *Flight from China.* New York: Macmillan.

Bourdieu, Pierre. 1977. *Outline of a Theory of Practice,* translated by R. Nice. Cambridge: Cambridge University Press.

Brascope, Simon. 1994. "Strategic Adaptations: Native Aesthetics Through Time and Space." Pp. 93–96 in *Native American Expressive Culture: Museum of the American Indian.* Ithaca, NY: Cornell University, Akwe:kon Press/ NMAI (distributed by Fulcrum Press).

Bryan, Dominic. 2000. *Orange Parades.* Sterling, VA: Pluto.

Bullock, Alan. 1992. *Hitler and Stalin: Parallel Lives.* New York: Alfred A. Knopf.

Burden, Hamilton T. 1967. *The Nuremberg Party Rallies: 1923–39.* New York: Frederick A. Praeger.

CAIN (Conflict Archive on the Internet). "Parades and Marches—Background Information on the Main Parading Organizations." Available online at www.cain.ulst.ac.uk/issues/parade/organis.htm. Material is added to this site on a regular basis (specific dates not provided).

Call, Dwight William. 1991. "The Sacred Tree Flowers: The Continuation of Culture within the Niches on Cheyenne River Reservation (Lakota Sioux, South Dakota)." Ph.D. diss., Drew University.

Camic, Charles. 1986. "The Matter of Habit." *American Journal of Sociology* 91:1039–87.

Campan, Mme. 1825. *Lettres de deux amies, ou correspondance entre deux élèves d'Ecouen.* Paris: Baudouin Frères.

Cappelle, Marie. 1867. *Mémoires.* Brussels: Hauman.

Cecil, Rosanne. 1993. "The Marching Season in Northern Ireland: An Expression of Politico-Religious Identity." Pp. 146–66 in *Inside European Identities,* edited by S. MacDonald. Oxford: Berg.

Cédoz, L'Abbé F.-M.-Th. 1891. *Un couvent de religieuses anglaises à Paris de 1634 à 1884, ouvrage.* Paris: Gallimard.

Cesbron, Gilbert. 1948. *Notre prison est un royaume.* Paris: Le Cercle du Livre de France.

Champagne, Duane, ed. 1999. *Contemporary Native American Cultural Issues.* Walnut Creek, CA: AltaMira.

Chew, Sing C., and J. David Knottnerus, eds. 2002. *Structure, Culture and History: Recent Issues in Social Theory.* Lanham, MD: Rowman & Littlefield.

Cocteau, Jean. 1925. *Les enfants terribles.* Paris: Bernard Grasset.

Cohen, Abner. 1969. "Political Anthropology: The Analysis of the Symbolism of Power Relations." *Man* 4:215–35.

———. 1974. *Two Dimensional Man: An Essay on the Anthropology of Power and Symbolism in Complex Society.* Berkeley: University of California Press.

———. 1979. "Political Symbolism." Pp. 87–113 in *Annual Review of Anthropology,* vol. 8, edited by Bernard J. Siegel, Alan R. Beals, and Stephen A. Tyler. Palo Alto, CA: Annual Reviews.

Collins, Randall. 1975. *Conflict Sociology: Toward an Explanatory Science.* New York: Academic Press.

———. 1981. "On the Microfoundations of Macrosociology." *American Journal of Sociology* 86:984–1014.

———. 1984. "Statistics versus Words." Pp. 329–62 in *Sociological Theory 1984,* edited by Randall Collins. San Francisco: Jossey-Bass.

———. 1987. "Interaction Ritual Chains, Power and Property: The Micro-Macro Connection as an Empirically Based Theoretical Problem." Pp. 193–206 in *The Micro-Macro Link,* edited by Jeffrey C. Alexander, Bernard Giesen, Richard Munch, Neil J. Smelser. Berkeley: University of California Press.

———. 2004. *Interaction Ritual Chains.* Princeton, NJ: Princeton University Press.

Connerton, Paul. 1989. *How Societies Remember.* Cambridge: Cambridge University Press.

Cook, Karen S., ed. 1987. *Social Exchange Theory.* Newbury Park, CA: Sage.

Cornell, Stephen. 1988. *The Return of the Native: American Indian Political Resurgence.* New York: Oxford University Press.

Crubellier, Maurice. 1979. *L'enfance et la jeunesse dans la société française, 1800–1950.* Paris: Librairie Armand Colin.

Darby, John. 1997. *Scorpions in a Bottle: Conflicting Cultures in Northern Ireland.* London: Minority Rights Publications.

Dasari, Adam, and J. David Knottnerus. In progress. "The Production of Ritual and Its Effects in an American Asian Indian Community."

Davanture, Maurice. 1975. *La jeunesse de Maurice Barrès (1862–1888).* Paris: Librairie Honoré Champion.

De Certeau, Michel. 1988. *The Practice of Everyday Life.* Berkeley: University of California Press.

Diephouse, David J. 1983. "The Triumph of Hitler's Will." Pp. 51-76 in *The Cult of Power: Dictators in the Twentieth Century,* edited by J. Held. Boulder, CO: East European Monographs (distributed by Columbia University Press, New York).

DiMaggio, Paul J., and Walter W. Powell. 1983. "The Iron Cage Revisited: Institutional Isomorphism and Collective Rationality in Organizational Fields." *American Sociological Review* 48:147–60.

Douglas, Mary. 1970. *Natural Symbols.* New York: Vintage.

Duhamel, Georges. 1949. *Inventaire de l'abîme, 1884–1901: Lumières sur ma vie I.* Paris: Mercure de France.

Durkheim, Emile. [1915] 1965. *The Elementary Forms of the Religious Life.* New York: Free Press.

———. [1897] 1966. *Suicide: A Study of Sociology.* New York: Free Press.

Edelman, Murray. 1995. *From Art to Politics: How Artistic Creations Shape Political Conceptions.* Chicago: University of Chicago Press.

Edwards, Jennifer, and J. David Knottnerus. 2007. "The Orange Order: Strategic Ritualization and Its Organizational Antecedents." *International Journal of Contemporary Sociology* 44:179–99.

———. 2009. "Exchange, Conflict and Coercion: The Ritual Dynamics of the Notting Hill Carnival Past and Present." Presented at the annual meeting of the American Association of Behavioral and Social Sciences, Las Vegas.

———. 2010. "The Orange Order: Parades, Other Rituals, and Their Outcomes." *Sociological Focus* 43:1–23.

Ehrenreich, Barbara. 2006. *Dancing in the Streets: A History of Collective Joy*. New York: Metropolitan Books (Henry Holt and Company).

Engelstad, Fredrik, ed. 2003. *Comparative Studies of Culture and Power: Comparative Social Research,* vol. 21. Oxford: Elsevier JAI.

Farnell, B. 1999. "Moving Bodies, Acting Selves." *Annual Review of Anthropology* 28:341–73.

Farrell, Michael. 1980. *Northern Ireland: The Orange State*. London: Pluto.

Fest, Joachim C. 1973. *Hitler*. New York: Harcourt Brace Jovanovich.

Fiske, Susan T., and Shelley E. Taylor. 1991. *Social Cognition*. 2nd ed. New York: McGraw-Hill.

Fixico, Donald. 2000. *The Urban Indian Experience in America*. Albuquerque: University of New Mexico Press.

Forgas, Joseph P. 1981. *Social Cognition: Perspectives on Everyday Understanding*. London: Academic Press.

Foucault, Michel. 1975. *Discipline and Punish: The Birth of the Prison*. New York: Vintage.

Fournier, Alain. 1966. *Meaulnes: The Lost Domain,* translated by Sandra Morris. London: Blackie & Son.

Friar, Ralph E., and N. Friar. 1972. *The Only Good Indian: The Hollywood Gospel*. New York: Drama Book Specialists.

Friedland, Roger, and Robert R. Alford. 1991. "Bringing Society Back In: Symbols, Practices, and Institutional Contradictions." Pp. 232–63 in *The New Institutionalism in Organizational Analysis,* edited by Walter W. Powell and Paul J. DiMaggio. Chicago: University of Chicago Press.

Gamble, John I. 1952. "Changing Patterns of Kiowa Indian Dances." *International Congress of Americanists,* Proceedings 29:94–104.

Gamm, Hans-Jochen. 1962. *Der braune Kult: Das Dritte Reich und seine Ersatzreligion*. Hamburg: Rutten & Loening.

Garfinkel, Harold. 1967. *Studies in Ethnomethodology*. Englewood Cliffs, NJ: Prentice-Hall.

Gehrke, Nathalie J. 1979. "Rituals of the Hidden Curriculum." Pp. 103–27 in *Children in Time and Space,* edited by Kaoru Yamamoto. New York: Teachers College Press.

Gelo, Daniel Joseph. 1986. "Comanche Belief and Ritual" (American Indian Religion, Oklahoma, Structuralism, Ethnography). Ph.D. diss., Rutgers State University.

Gerbod, Paul. 1968. *La vie quotidienne dans les lycées et collèges au XIXe siècle*. Paris: Hachette.

Giddens, Anthony. 1979. *Central Problems in Social Theory*. Berkeley: University of California Press.

———. 1984. *The Constitution of Society*. Berkeley: University of California Press.

Gide, André. 1925. *Les faux-monnayeurs*. Paris: Gallimard, Le Livre de Poche.

———. 1958. *Si le grain ne meurt*. Paris: Gallimard.

Goffman, Erving. 1961. *Asylums: Essays on the Social Situation of Mental Patients and Other Inmates*. Garden City, NY: Anchor Books.

———. 1967. *Interaction Ritual: Essays on Face-to-Face Behavior*. Garden City, NY: Anchor Books.

Grafteaux, Serge. 1985. *Mémé Santerre: A French Woman of the People*, translated by Louise A. Tilly and Kathryn L. Tilly. New York: Schocken Books.

"The Grand Orange Lodge of Ireland." Available online at www.grandorangelodge.co.uk. Updated with press releases on a regular basis (specific dates not provided).

Guan, Jian, and J. David Knottnerus. 1999. "A Structural Ritualization Analysis of the Process of Acculturation and Marginalization of Chinese Americans." *Humboldt Journal of Social Relations* 25:43–95.

———. 2002. "A Structural Ritualization Analysis of the Process of Acculturation and Marginalization of Chinese Americans." *World Ethno-National Studies* 1:68–81 (Chinese academic journal; translation of 1999 article).

———. 2006. "Chinatown under Siege: Community Protest and Structural Ritualization Theory." *Humboldt Journal of Social Relations* 30:5–52.

Haddick-Flynn, Kevin. 1999. *Orangeism: The Making of a Tradition*. New York: St. Martin's.

Harris, Scott R. 2000. "The Social Construction of Equality in Everyday Life." *Human Studies* 23:371–93.

Hays, Sharon. 1994. "Structure and Agency and the Sticky Problem of Culture." *Sociological Theory* 12:57–72.

Herling, Gustav. 1951. *A World Apart*, translated by Joseph Marek. New York: Roy Publishers.

Herman, Stewart W., Jr. 1943. *It's Your Souls We Want*. New York: Harper & Brothers.

Hertzberg, Hazel. 1971. *The Search for an American Indian Identity: Modern Pan-Indian Movements*. Syracuse, NY: Syracuse University Press.

Herzstein, Robert Edwin. 1978. *The War That Hitler Won: The Most Infamous Propaganda Campaign in History*. New York: G. P. Putnam's Sons.

Heth, Charlotte, ed. 1992. *Native American Dance: Ceremonies and Social Traditions*. Washington, DC: National Museum of the American Indian, Smithsonian Institution with Starwood Publishing.

Hobsbawm, Eric, and Terence Ranger, eds. 1983. *The Invention of Tradition*. Cambridge: Cambridge University Press.

Hocart, A. M. [1950] 1968. *Caste: A Comparative Study*. New York: Russell & Russell.

Hollander, Jocelyn A., and Judith A. Howard. 2000. "Social Psychological Theories on Social Inequality." *Social Psychology Quarterly* 63:338–51.

Holmes, Thomas, and Richard Rahe. 1967. "Social Readjustment Rating Scale." *Journal of Psychosomatic Research* 11:214.

Howard, Judith A. 1994. "A Social Cognitive Conception of Social Structure." *Social Psychology Quarterly* (special issue) 57:210–27.

———. 1995. "Social Cognition." Pp. 90–117 in *Sociological Perspectives on Social Psychology,* edited by Karen S. Cook, Gary Allen Fine, and James S. House. Boston: Allyn & Bacon.

Hunt, Lynn. 1984. *Politics, Culture, and Class in the French Revolution.* Berkeley: University of California Press.

———. 1988. "The Sacred and the French Revolution." Pp. 25–43 in *Durkheimian Sociology: Cultural Studies,* edited by Jeffrey C. Alexander. Cambridge: Cambridge University Press.

Iverson, Peter. 1999. *Riders of the West.* Vancouver: Greystone Books.

Jäckel, Eberhard. 1972. *Hitler's Weltanschauung: A Blueprint for Power,* translated by Herbert Arnold. Middletown, CT: Wesleyan University Press.

Jammes, Francis. 1971. *Mémoires.* Paris: Mercure de France.

Jarman, Neil. 1996. "Violent Men, Violent Land: Dramatizing the Troubles and Landscape of Ulster." *Journal of Material Culture* 1:39–61.

———. 1997. *Material Conflicts: Parades and Visual Displays in Northern Ireland.* Oxford: Berg.

Kapferer, Judith L. 1981. "Socialization and the Symbolic Order of the School." *Anthropology and Education Quarterly* 12:258–74.

Kavanagh, Thomas W. 1992. "Southern Plains Dance: Tradition and Dynamics." Pp. 105–23 in *Native American Dance: Ceremonies and Social Traditions,* edited by Charlotte Heth. Washington, DC: National Museum of the American Indian, Smithsonian Institution, with Starwood Publishing.

Kenney, Mary Catherine. 1991. "Neighborhoods and Parades: The Social and Symbolic Organization of Conflict in Northern Ireland." Ph.D. diss., University of Michigan.

Kertzer, David I. 1988. *Ritual, Politics, and Power.* New Haven, CT: Yale University Press.

Kincaid, Harold. 1996. *Philosophical Foundations of the Social Sciences.* New York: Cambridge University Press.

Kitagawa, Daisuke. 1967. *Issei and Nisei: The Internment Years.* New York: Seabury.

Knottnerus, J. David. 1997. "The Theory of Structural Ritualization." Pp. 257–79 in *Advances in Group Processes,* vol. 14, edited by Barry Markovsky, Michael J. Lovaglia, and Lisa Troyer. Greenwich, CT: JAI Press.

———. 1999. "Status Structures and Ritualized Relations in the Slave Plantation System." Pp. 137–47 in *Plantation Society and Race Relations: The Origins of Inequality,* edited by Thomas J. Durant, Jr., and J. David Knottnerus. Westport, CT: Praeger.

———. 2002. "Agency, Structure and Deritualization: A Comparative Investigation of Extreme Disruptions of Social Order." Pp. 85–106 in *Structure, Culture and History: Recent Issues in Social Theory,* edited by Sing C. Chew and J. David Knottnerus. Lanham, MD: Rowman & Littlefield.

———. 2005. "The Need for Theory and the Value of Cooperation: Disruption and Deritualization." Presidential Address, Mid-South Sociological Association, Baton Rouge, 2003. *Sociological Spectrum* 25:5–19.

———. 2007. "Social Psychology." Pp. 153–60 in *21st Century Sociology: A Reference Handbook,* vol. 1, edited by Clifton D. Bryant and Dennis L. Peck. Thousand Oaks, CA: Sage Publications.

———. 2009. "Structural Ritualization Theory: Application and Change." Pp. 70–84 in *Bureaucratic Culture and Escalating World Problems: Advancing the Sociological Imagination,* edited by J. David Knottnerus and Bernard Phillips. Boulder, CO: Paradigm Publishers.

———. Forthcoming. "Collective Events, Rituals and Emotions." In *Advances in Group Processes,* vol. 27, edited by Shane R. Thye and Edward J. Lawler. Bingley, United Kingdom: Emerald Group Publishing Limited.

Knottnerus, J. David, and Phyllis E. Berry. 2002. "Spartan Society: Structural Ritualization in an Ancient Social System." *Humboldt Journal of Social Relations* 27:1–42.

Knottnerus, J. David, and David G. LoConto. 2003. "Strategic Ritualization and Ethnicity: A Typology and Analysis of Ritual Enactments in an Italian American Community." *Sociological Spectrum* 23:425–61.

Knottnerus, J. David, David L. Monk, and Edward Jones. 1999. "The Slave Plantation System from a Total Institution Perspective." Pp. 17–27 in *Plantation Society and Race Relations: The Origins of Inequality,* edited by Thomas J. Durant, Jr., and J. David Knottnerus. Westport, CT: Praeger.

Knottnerus, J. David, and Bernard Phillips, eds. 2009. *Bureaucratic Culture and Escalating World Problems: Advancing the Sociological Imagination.* Boulder, CO: Paradigm Publishers.

Knottnerus, J. David, and Christopher Prendergast, eds. 1994. *Recent Developments in the Theory of Social Structure.* Current Perspectives in Social Theory. Greenwich, CT: JAI Press.

Knottnerus, J. David, Jason S. Ulsperger, Summer Cummins, and Elaina Osteen. 2006. "Exposing Enron: Media Representations of Ritualized Deviance in Corporate Culture." *Crime, Media, Culture* 2:177–95.

Knottnerus, J. David, and Frédérique Van de Poel-Knottnerus. 1999. *The Social Worlds of Male and Female Children in the Nineteenth Century French Educational System: Youth, Rituals and Elites.* Lewiston, NY: Edwin Mellen.

Kracht, Benjamin R. 1994. "Kiowa Powwows: Continuity in Ritual Practice." *American Indian Quarterly* 3:321–28.

La Forte, Robert S., Ronald E. Marcello, and Richard L. Himmel. 1994.

With Only the Will to Live: Accounts of Americans in Japanese Prison Camps 1941–1945. Wilmington, DE: Scholarly Resources Books.

Lane, Christel. 1981. *The Rites of Rulers: Ritual in Industrial Society—The Soviet Case.* Cambridge: Cambridge University Press.

Lassiter, Luke E. 1999. "Southwestern Oklahoma, the Gourd Dance, and 'Charlie Brown.'" Pp. 145–66 in *Contemporary Native American Cultural Issues,* edited by Duane Champagne. Walnut Creek, CA: AltaMira.

Lavergne, Joseph. 1903. *Correspondance de madame Julie Lavergne, deuxième partie, lettres de 1871 à 1886.* Paris: Taffin-Lefort.

Lawler, Edward J., and Rebecca Ford. 1993. "Metatheory and Friendly Competition in Theory Growth: The Case of Power Processes in Bargaining." Pp. 172–210 in *Theoretical Research Programs: Studies in the Growth of Theory,* edited by Joseph Berger and Morris Zelditch, Jr. Stanford, CA: Stanford University Press.

Lerch, Patricia Barker, and Susan Bullers. 1996. "Powwows as Identity Markers: Traditional or Pan-Indian?" *Human Organization* 55:390–95.

Lowenthal, Leo. 1961. *Literature, Popular Culture, and Society.* Englewood Cliffs, NJ: Prentice-Hall.

———. 1986. *Literature and the Image of Man. Communication in Society,* vol. 2. New Brunswick, NJ: Transaction Books.

Lui, Robert K. 2000. "A Black Hills Powwow." *Ornament* 4:22–24.

Mach, Zdzislaw. 1992. "Continuity and Change in Political Ritual: May Day in Poland." Pp. 43–61 in *Revitalizing European Rituals,* edited by Jeremy Boissevain. London: Routledge.

———. 1993. *Symbols, Conflict, and Identity: Essays in Political Anthropology.* Albany: State University of New York Press.

Malraux, Clara. 1967. *Clara Malraux: Memoirs,* translated by Patrick O'Brian. New York: Farrar, Straus & Giroux.

Markus, H., and Robert B. Zajonc. 1985. "The Cognitive Perspective in Social Psychology." Pp. 137–230 in *Handbook of Social Psychology,* vol. 1, *Theory and Method,* 3rd ed., edited by Gardner Lindzey and Elliot Aronson. New York: Random House.

Marthelet, Pierre. 1993. *Une dynastie d'écolâtres.* Paris: Editions Tiresias.

Marx, Karl. 1967. *Capital.* New York: International Publishers.

———. 1970. *A Contribution to the Critique of Political Economy,* translated by S. W. Ryazanskaya. New York: International Publishers.

Mason, James D., and J. David Knottnerus. 2009. "The Application of Structural Ritualization Theory to Polar Expeditions." Presented at the 104th annual meeting of the American Sociological Association, San Francisco.

Mathews, Lita. 1999. "The Native American Powwow: A Contemporary Authentication of a Cultural Artifact." Ph.D. diss., University of New Mexico.

Mattern, Mark. 1999. "The Powwow as a Public Arena for Negotiating Unity and Diversity in American Indian Life." Pp. 129–43 in *Contemporary Native American Cultural Issues,* edited by Duane Champagne. Walnut Creek, CA: AltaMira.

Maurois, André. 1953. *Lelia: The Life of George Sand,* translated by Gerard Hopkins. New York: Harper & Brothers.

Maynard, Douglas W. 1996. "On 'Realization' in Everyday Life: The Forecasting of Bad News as a Social Relation." *American Sociological Review* 61:109–31.

McNickle, D'Arcy. [1973] 1993. *Native American Tribalism: Indian Survivals and Renewals.* Oxford: New York.

Meyer, John W., and Michael T. Hannan. 1979. *National Development and the World System: Educational, Economic, and Political Change, 1950–1970.* Chicago: University of Chicago Press.

Meyer, John W., and Brian Rowan. 1977. "Institutionalized Organizations: Formal Structure as Myth and Ceremony." *American Journal of Sociology* 83:340–63.

Mills, C. Wright. 1959. *The Sociological Imagination.* New York: Oxford University Press.

Minton, Carol, and J. David Knottnerus. 2008. "Ritualized Duties: The Social Construction of Gender Inequality in Malawi." *International Review of Modern Sociology* 34:181–210.

Mitra, Aditi, and J. David Knottnerus. 2004. "Royal Women in Ancient India: The Ritualization of Inequality in a Patriarchal Social Order." *International Journal of Contemporary Sociology* 41:215–31.

———. 2008. "Sacrificing Women: A Study of Ritualized Practices Among Women Volunteers in India." *Voluntas: International Journal of Voluntary and Nonprofit Organizations* 19:242–67.

Momaday, N. Scott. 1975. "To the Singing, to the Drums." *Natural History* 134:38–45.

Moore, Barrington, Jr. 2000. *Moral Purity and Persecution in History.* Princeton, NJ: Princeton University Press.

Mosse, George L. 1964. *The Crisis of German Ideology: Intellectual Origins of the Third Reich.* London. Grosset & Dunlap.

———. 1975. *The Nationalization of the Masses: Political Symbolism and Mass Movements from the Napoleonic Wars Through the Third Reich.* New York: Meridian, New American Library.

Nadel, S. F. 1954. *Nupe Religion.* Glencoe, IL: Free Press.

———. 1957. *The Theory of Social Structure.* Glencoe, IL: Free Press.

Nagel, Joane. 1982. "The Political Mobilization of Native Americans." *Social Science Journal* 19:37–46.

———. 1997. *American Indian Ethnic Renewal: Red Power and the Resurgence of Identity and Culture.* New York: Oxford University Press.

Nisbet, Robert A. 1966. *The Sociological Tradition.* New York: Basic Books.

O'Malley, Padraig. 1995. *Religion and Conflict: The Case of Northern Ireland.* Boston: John W. McCormack Institute of Public Affairs.

Ozouf, Mona. 1975. "Space and Time in the Festivals of the French Revolution." *Comparative Studies in Society and History* 17:372–84.

Pagnol, Marcel. 1977. *Livre de la nature, souvenirs d'enfance, 1. La gloire de mon père, 2. Le château de ma mère. Oeuvres complètes de Marcel Pagnol, II.* Paris: Editions du Club de l'Honnête Homme.

Peterson, R. 1999. "From Politics to Powwows: Boye Ladd Serves His People Dancing." *Aboriginal Voices: A Native American News Magazine* 6 (July-August):12–13.

Petrone, Karen. 2000. *Life Has Become More Joyous, Comrades: Celebrations in the Time of Stalin.* Bloomington: Indiana University Press.

Phillips, Bernard. 2001. *Beyond Sociology's Tower of Babel: Reconstructing the Scientific Method.* New York: Aldine de Gruyter.

———. 2007. *Understanding Terrorism: Building on the Sociological Imagination.* Boulder, CO: Paradigm Publishers.

———. 2009. *Armageddon or Evolution? The Scientific Method and Escalating World Problems.* Boulder, CO: Paradigm Publishers.

Phillips, Bernard, and David Christner. Forthcoming. *Saving Society: Beyond Bureaucratic Barriers to Real Solutions.* Boulder, CO: Paradigm Publishers.

Phillips, Bernard, and Louis Johnston. 2007. *The Invisible Crisis of Contemporary Society: Reconstructing Sociology's Fundamental Assumptions.* Boulder, CO: Paradigm Publishers.

Phillips, Bernard, Harold Kincaid, and Thomas J. Scheff, eds. 2002. *Toward a Sociological Imagination: Bridging Specialized Fields.* Lanham, MD: University Press of America.

Powell, Walter W., and Paul J. DiMaggio. 1991. *The New Institutionalism in Organizational Analysis.* Chicago: University of Chicago Press.

Powers, William K. 1990. *War Dance: Plains Indian Musical Performance.* Tucson: University of Arizona Press.

Prendergast, Christopher, and J. David Knottnerus. 1993. "The New Studies in Social Organization: Overcoming the Astructural Bias." Pp. 158–85 in *Interactionism: Exposition and Critique,* 3rd ed., edited by Larry T. Reynolds. Dix Hills, NY: General Hall.

———. 1994. "Recent Developments in the Theory of Social Structure: Introduction and Overview." Pp. 1–26 in *Recent Developments in the Theory of Social Structure,* edited by J. David Knottnerus and Christopher Prendergast. Current Perspectives in Social Theory. Greenwich, CT: JAI Press.

Rappaport, Roy A. 1979. *Ecology, Meaning and Religion.* Berkeley, CA: North Atlantic Books.

———. 1999. *Ritual and Religion in the Making of Humanity.* Cambridge: Cambridge University Press.

Rivaux, The Abbé. 1887. *Life of Rev. Mother St. John Fontbonne, Foundress and First Superior-General of the Congregation of the Sisters of St. Joseph,* translated by Honorary Canon. New York: Benziger Brothers.

Sanchez, Victoria Eugenie. 1995. "'As Long as We Dance, We Shall Know Who We Are': A Study of Off-Reservation Traditional Inter-Tribal Powwows in Central Ohio." Ph.D. diss., Ohio State University.

Sand, George. [1854] 1970. *Oeuvres autobiographiques, I. Histoire de ma vie.* Paris: Editions Gallimard, Bibliothèque de la Pléiade.

———. 1979. *My Life: By George Sand,* translated by Dan Hofstadter. New York: Harper & Row.

Sarabia, Daniel, and J. David Knottnerus. 2009. "Ecological Stress and Deritualization in East Asia: Ritual Practices during Dark Age Phases." *International Journal of Sociology and Anthropology* (open access online journal) 1, no. 1 (May): 12–25. Available online at www.academicjournals.org/IJSA/contents/2009cont/May.htm.

Scheff, Thomas J. 1997. *Emotions, the Social Bond, and Human Reality: Part/Whole Analysis.* Cambridge: Cambridge University Press.

———. 2006. *Goffman Unbound! A New Paradigm for Social Science.* Boulder, CO: Paradigm Publishers.

Schwalbe, Michael, Sandra Godwin, Daphne Holden, Douglas Schrock, Shealy Thompson, and Michelle Wolkomir. 2000. "Generic Processes in the Reproduction of Inequality: An Interactionist Analysis." *Social Forces* 79:419–52.

Sell, Jane. 2007. "The Sociology of Group Dynamics." Pp. 171–78 in *21st Century Sociology: A Reference Handbook,* vol. 1, edited by Clifton D. Bryant and Dennis L. Peck. Thousand Oaks, CA: Sage Publications.

Sell, Jane, J. David Knottnerus, and Christina Adcock. Unpublished manuscript. "Disruptions in Task Groups: When Does Disruption Lead to Disorder?"

Sell, Jane, J. David Knottnerus, Christopher Ellison, and Heather Mundt. 2000. "Reproducing Social Structure in Task Groups: The Role of Structural Ritualization." *Social Forces* 79:453–75.

Sewell, William H. 1992. "A Theory of Structure: Duality, Agency, and Transformation." *American Journal of Sociology* 98:1–29.

Shils, Edward, and M. Young. 1953. "The Meaning of the Coronation." *Sociological Review* 1:63–81.

Simmel, Georg. 1906. "The Sociology of Secrecy and of Secret Societies." *American Journal of Sociology* 11:441–98.

———. 1950. *The Sociology of Georg Simmel,* edited by Kurt H. Wolff. Glencoe, IL: Free Press.

Simpson, Joseph M., J. David Knottnerus, and Michael J. Stern. 2009. "Community and Ritual in Virtual Worlds." Presented at the annual meeting of the Southern Sociological Society, New Orleans.

Smith, Dean Howard. 2000. *Modern Tribal Development: Paths to Self-Sufficiency and Cultural Integrity in Indian Country.* Walnut Creek, CA: AltaMitra.

Snipp, C. Matthew. 1996. "The First Americans: American Indians." Pp. 131–43 in *Majority and Minority: The Dynamics of Race and Ethnicity in American Life,* 6th ed., edited by Norman R. Yetman. Boston: Allyn & Bacon.

Steele, C. Hoy. 1973. "Urban Indian Identity in Kansas: Some Implications from Research." *Proceedings of the American Ethnological Society,* pp. 167–78.

Stern, Daniel (Marie d'Agoult). 1877. *Mes souvenirs, 1806–1833.* Paris: Calmann-Lévy, Editeur.

Stinchcombe, Arthur L. 1973. "Formal Organizations." Pp. 23–66 in *Sociology,* 2nd ed., edited by Neil J. Smelser. New York: Wiley.

Sylvère, Antoine. 1980. *Toinou: Le cri d'un enfant auvergnat, pays d'Ambert.* Paris: Plon.

Szmatka, Jacek, John Skvoretz, and Joseph Berger, eds. 1997. *Status, Network, and Structure.* Stanford, CA: Stanford University Press.

Taylor, Simon. 1981. "Symbol and Ritual under National Socialism." *British Journal of Sociology* 32:504–20.

———. 1985. *Prelude to Genocide: Nazi Ideology and the Struggle for Power.* New York: St. Martin's.

Thornburg, P. Alex, and J. David Knottnerus. 2008. "Consumer Ritualized Symbolic Practices: A Theory of Religious Commodification." Presented at the 103rd annual meeting of the American Sociological Association, Boston.

Thornburg, P. Alex, J. David Knottnerus, and Gary R. Webb. 2007. "Disaster and Deritualization: A Re-interpretation of Findings from Early Disaster Research." *The Social Science Journal* 44:161–66.

———. 2008. "Ritual and Disruption: Insights from Early Disaster Research." *International Journal of Sociological Research* 1:91–109.

Todd, Jennifer. 1994. "History and Structure in Loyalist Ideology: The Possibilities of Ideological Change." *Irish Journal of Sociology* 4:67–79.

Troyer, Lisa, and Steven D. Silver. 1999. "Institutional Logics and Group Environments: Toward an Open System Perspective on Group Processes." Pp. 219–52 in *Advances in Group Processes,* vol. 16, edited by Shane R. Thye, Edward J. Lawler, Michael Macy, and Henry A. Walker. Greenwich, CT: JAI Press.

Turin, Yvonne. 1989. *Femmes et religieuses au XIXe siècle, le féminisme "en religion."* Paris: Nouvelle Cité.

Turner, Jonathan. 2003. *The Structure of Sociological Theory,* 7th ed. Belmont, CA: Wadsworth.

Turner, Ralph H. 1991. "The Use and Misuse of Rational Models in Collective Behavior and Social Psychology." *European Journal of Sociology* 32:84–108.

Turner, Victor W. 1967. *The Forest of Symbols: Aspects of Ndembu Ritual.* Ithaca, NY: Cornell University Press.

Ulsperger, Jason S., and J. David Knottnerus. 2006. "Enron: Organizational Rituals as Deviance." Pp. 279–82 in *Readings in Deviant Behavior,* 4th ed., edited by Alex Thio and Thomas C. Calhoun. Boston: Allyn & Bacon.

———. 2007. "Long-term Care Workers and Bureaucracy: The Occupational Ritualization of Maltreatment in Nursing Homes and Recommended Policies." *Journal of Applied Social Science* 1:52–70.

———. 2008. "The Social Dynamics of Elder Care: Rituals of Bureaucracy and Physical Neglect in Nursing Homes." *Sociological Spectrum* 28:357–88.

———. 2009a. "Illusions of Affection: Bureaucracy and Emotional Neglect in Nursing Homes." *Humanity and Society* 33:238–59.

———. 2009b. "Institutionalized Elder Abuse: Bureaucratic Ritualization and Transformation of Physical Neglect in Nursing Homes." Pp. 134–55 in *Bureaucratic Culture and Escalating World Problems: Advancing the Sociological Imagination,* edited by J. David Knottnerus and Bernard Phillips. Boulder, CO: Paradigm Publishers.

———. Forthcoming. *Elder Care Catastrophe: Rituals of Abuse in Nursing Homes—And What You Can Do About It.* Boulder, CO: Paradigm Publishers.

Unger, Aryeh L. 1974. *The Totalitarian Party: Party and People in Nazi Germany and Soviet Russia.* London: Cambridge University Press.

Utter, Jack. 1992. *Wounded Knee and the Ghost Dance Tragedy.* Lake Ann, MI: National Woodlands Publishing Company.

Van de Poel-Knottnerus, Frédérique, and J. David Knottnerus. 2002. *Literary Narratives on the Nineteenth and Early Twentieth-Century French Elite Educational System: Rituals and Total Institutions.* Lewiston, NY: Edwin Mellen.

Varner, Monica K., and J. David Knottnerus. 2002. "Civility, Rituals and Exclusion: The Emergence of American Golf During the Late Nineteenth and Early Twentieth Centuries." *Sociological Inquiry* 72:426–41.

Vaughan, Diane. 1996. *The Challenger Launch Decision: Risky Technology, Culture, and Deviance at NASA.* Chicago: University of Chicago Press.

Vennum, Thomas, Jr. 1989. "The Changing Role of Women in Ojibway Music History." Pp. 13–25 in *Women in North American Indian Music,* edited by Richard Keeling. Special Series no. 6. Bloomington, IN: Society for Ethnomusicology.

Vondung, Klaus. 1971. *Magie und Manipulation: Ideologischer Kult und politische Religione des Nationalsozialismus.* Göttingen: Vandenhoeck & Ruprecht.

Wagener, Françoise. 1999. *L'Impératrice Joséphine (1763–1814).* Paris: Flammarion.

Wallerstein, Immanuel. 1974. *The Modern World-System,* vol. 1, *Capitalist Agriculture and the Origins of the European World-Economy in the Sixteenth Century.* New York: Academic Press.

Warner, W. Lloyd. 1959. *The Living and the Dead: A Study of the Symbolic Life of Americans.* Yankee City Series. New Haven, CT: Yale University Press.

———. 1962. *American Life: Dream and Reality.* Rev. ed. Chicago: University of Chicago Press.

Warriner, Charles K. 1981. "Levels in the Study of Social Structure." Pp. 179–90 in *Continuities in Structural Inquiry,* edited by Peter M. Blau and Robert K. Merton. Beverly Hills, CA: Sage.

Weibel-Orlando, Joan. 1991. *Indian Country, L.A.: Maintaining Ethnic Community in Complex Society.* Champaign: University of Illinois Press.

Welch, David. 1993. *The Third Reich: Politics and Propaganda.* London: Routledge.

White, Debbie, and J. David Knottnerus. In progress. "Ritualization and Coping with HIV Infection."

Whitehorse, David. 1988. *Pow-wow: The Contemporary Pan-Indian Celebration.* Publications in American Indian Studies 5. San Diego: San Diego State University.

Whitmeyer, Joseph, and Karen Cook. 2002. "Social Structure and Social Exchange." Pp. 271–302 in *Structure, Culture, and History: Recent Issues in Social Theory,* edited by Sing Chew and J. David Knottnerus. Lanham, MD: Rowman & Littlefield.

Willer, David. 1999. *Network Exchange Theory.* Westport, CT: Praeger.

Willer, David, and Bo Anderson, eds. 1981. *Networks, Exchange and Coercion.* New York: Elsevier.

Wolf, Eric R. 1999. *Envisioning Power: Ideologies of Dominance and Crisis.* Berkeley: University of California Press.

Wu, Yanhong, and J. David Knottnerus. 2005. "Ritualized Daily Practices: A Study of Chinese 'Educated Youth.'" *Shehui* (*Society*) 6:167–85 (Chinese academic journal).

———. 2007. "The Origins of Ritualized Daily Practices: From Lei Feng's Diary to Educated Youth's Diaries." *Shehui* (*Society*) 1:98–119 (Chinese academic journal).

———. 2008. "Ritualized Daily Practices: A Study of Chinese 'Educated Youth.'" Pp. 340–48 in *Research Methods in Social Science,* by Qiu Liping. Chongqing: Chongqing University Press (reprint of 2005 article).

———. In progress. "China's Educated Youth: Rituals of Story-Telling During the Cultural Revolution."

Young, Frank W. 1994. "Durkheim and Development Theory." *Sociological Theory* 12:73–82.

Young, Gloria Alese. 1981. "Pow wow Power: Perspectives on Historic and Contemporary Intertribalism." Ph.D. diss., Indiana University.

Zeman, Z. A. B. 1964. *Nazi Propaganda.* London: Oxford University Press.

Zentner, Christian. 1997. "Hitler's Worldview." Pp. 426–30 in *The Encyclopedia*

of the Third Reich, edited by Christian Zentner and Friedemann Bedurftig. New York: Da Capo Press.
Zentner, Christian, and Friedemann Bedurftig, eds. 1997. *The Encyclopedia of the Third Reich.* New York: Da Capo Press.

Chapter 4

Videography of Native American Powwows

Cultural Celebration: Symposium on the American Indian Pow-wow. 1998. Produced by Tim Tall Chief. Northeastern State University, Oklahoma.
Discover Oklahoma—Episode 20. 1991. Oklahoma Dept of Tourism.
Fancy Dance, vol. 1, *Native American Dance Series.* 1997. Full Circle Communications.
Finding the Circle. 1989. American Indian Dance Theatre.
Indian Symposium Powwow. 1994. Produced by Jake Bryan. Northeastern State University, Oklahoma.
Inter-tribal Pow-wow. 1995. Produced by Jake Bryan, Kevin Chanate and Keri Jhane Bread. Northeastern State University, Oklahoma.
Into the Circle: An Introduction to Native American Powwows. 1992. Produced by Scott Swearingen and Sandy Rhoades. Videography by Scott Swearingen. Full Circle Communications.
Native American Heritage Celebration: Keepers of the Stories . . . Celebrating the Indian Spirit. 1999. Oklahoma Department of Human Services.
Native American Men's and Women's Dance Styles, vol. 2. *Native American Dance Series.* 1997. Full Circle Communications.
Pow Wow Nation Earth Travelers, Oklahoma City, Oklahoma, June 1998, Men's Finals: Men's Fancy Dance of Champions. Pow Wow Nation.

Chapter 5

French Authors

Alizon, Simone. 1996. *L'exercice de vivre.* Paris: Editions Stock.
Borwicz, Michel. 1996. *Ecrits des condamnés à mort sous l'occupation nazie (1939–1945).* Paris: Gallimard.
Comte, Hubert. 1987. *La force de la colère.* Paris: Editions Stock.
Delbo, Charlotte. 1970. *Auschwitz et après,* I. *Aucun de nous ne reviendra.* Paris: Editions de Minuit.
———. 1970. *Auschwitz et après,* II. *Une connaissance inutile.* Paris: Editions de Minuit.

Heilbronn, Max. 1989. *Galeries Lafayette, Buchenwald, Galeries Lafayette …* Paris: Economica.
Hénocque, Abbé G. 1947. *Les antres de la bête … Fresnes Buchenwald Dachau.* Paris: G. Durassie.
Holt, Willy. 1995. *Femmes en deuil sur un camion.* Paris: NiL éditions.
Kammerer, Jean. 1995. *Mémoire en liberté: La baraque des prêtres à Dachau.* Paris: Editions Brepols.
Leroy, Roger, Roger Linet, and Max Nevers. 1991. *1943–1945 La résistance en enfer.* Paris: Messidor.
Michel, Jean. 1985. *De l'enfer aux étoiles. Dora. Le temps de la nuit.* Paris: Plon.
Novac, Ana. 1996. *Les beaux jours de ma jeunesse: Journal.* Paris: Editions Balland.
Pons, Francisco. 1993. *Barbelés à Argelès et autour d'autres camps.* Paris: L'Harmattan.
Rosenfeld, Etienne. 1991. *De Drancy à ces camps dont on ne parle pas.* Paris: L'Harmattan.
Rovan, Joseph. 1993. *Contes de Dachau.* Paris: Seuil.
Simier, Lucienne. 1992. *Deux ans au bagne de Ravensbrück.* Maulevrier: Hérault-Editions.
Songy, Jacques. 1985. *Fortes impressions de Dachau.* Châlons-Sur-Marne: Union Républicaine.

Non-French Authors

Beattie, Edward W., Jr. 1946. *Diary of a Kriege.* New York: Thomas Y. Crowell.
Birenbaum, Halina. 1996. *Hope Is the Last to Die.* Armonk, NY: M. E. Sharpe.
Cohen, Elie A. 1973. *The Abyss: A Confession.* New York: W. W. Norton.
de Wolff, Sal. 2000. *Fifty Years of Silence.* Self-published.
Frankel-Zaltzman, Paula. 2003. *Haftling (Prisoner) no. 94771: Experiences in German Lagers (Camps) 1941–1945: Lithuania: Dvinsk, Riga, Kaiserwald; Germany: Stutthof, Tarin, Bromberg.* Memoirs of Holocaust Survivors in Canada, vol. 28. Montreal: The Concordia University Chair in Canadian Jewish Studies and the Montreal Institute for Genocide and Human Rights Studies.
Frankl, Viktor E. 1972. *Man's Search for Meaning: An Introduction to Logotherapy.* New York: Pocket Books.
Gill, Anton. 1988. *The Journey Back from Hell: An Oral History.* Conversations with Concentration Camp Survivors. New York: William Morrow.
Herzberg, Abel J. 1997. *Between Two Streams: A Diary from Bergen-Belsen.* London: I. B. Tauris.
Huppert, Hilde. 1979. *Hand-in-Hand with Tommy: A Personal Report, 1939–1945.*

Self-published. (Translated into English by Yael Chaver, from the Hebrew version of Shmuel Tommy Huppert. This version appeared in 1978 by "Moreshet" and "Sifriat Poalim," Tel-Aviv.)

Ka-Tzetnik 135633. 1963. *Atrocity*. New York: Lyle Stuart.

Kielar, Wieslaw. 1980. *Anus Mundi: 1500 Days in Auschwitz/Birkenau*. New York: Times Books.

Korn, Abraham. 1995. *Abe's Story: A Holocaust Memoir*. Atlanta, GA: Longstreet.

Lengyel, Olga. 1947. *Five Chimneys: The Story of Auschwitz*. Chicago: Ziff-Davis.

Levi, Primo. 1986. *Survival in Auschwitz*; and, *The Reawakening: Two Memoirs*. New York: Summit Books.

Matzner, David. 1994. *The Muselmann: The Diary of a Jewish Slave Laborer*. Hoboken, NJ: KTAV.

Mechanicus, Philip. 1964. *Year of Fear: A Jewish Prisoner Waits for Auschwitz*. New York: Hawthorn Books.

Muller, Filip. 1979. *Eyewitness Auschwitz: Three Years in the Gas Chambers*. New York: Stein and Day.

Newman, Judith Sternberg. 1963. *In the Hell of Auschwitz: The Wartime Memoirs*. New York: Exposition.

Nomberg-Przytyk, Sara. 1985. *Auschwitz: True Tales from a Grotesque Land*. Chapel Hill: University of North Carolina Press.

Nyiszli, Dr. Miklos. 1980. *Auschwitz: A Doctor's Eye-witness Account*. London: Mayflower Book, Granada.

Perl, Dr. Gisella. 1984. *I Was a Doctor in Auschwitz*. Salem, NH: Ayer.

Rosenberg, Otto. 1999. *A Gypsy in Auschwitz* (as told to Ulrich Enzenberger). London: London House.

Rosicki, Tadeusz. 1984. *My Fifty Months in Nazi Camps: The Message of Survival*, translated by Maria T. Rosicki. N.p.

Sagan, S. J. 1982. *Food Carriers Out!* Self-published.

Szmaglewska, Seweryna. 1947. *Smoke over Birkenau*. New York: Henry Holt.

Weiss, Reska. 1961. *Journey Through Hell: A Woman's Account of Her Experiences at the Hands of the Nazis*. London: Vallentine, Mitchell.

Wiesel, Elie. 1982. *Night*. New York: Bantam.

Winkler, Ernst. 1942. *Four Years of Nazi Torture*. New York: D. Appleton-Century.

Secondary Source

Amery, Jean. 1980. *At the Mind's Limits: Contemplations by a Survivor on Auschwitz and Its Realities*. Bloomington: Indiana University Press.

Index

About the Author

J. David Knottnerus is Professor of Sociology at Oklahoma State University. He earned an undergraduate degree in sociology from Beloit College and an M.A. and a Ph.D. in sociology from Southern Illinois University–Carbondale. Knottnerus has published in the areas of social theory, social psychology, group processes, social structure and inequality, and collective behavior/social movements. Most of his efforts in recent years have focused on the development of structural ritualization theory and research. He is currently the coeditor—with Bernard Phillips—of the series Advancing the Sociological Imagination, with Paradigm Publishers. His most recent books are, coedited with Bernard Phillips, *Bureaucratic Culture and Escalating World Problems: Advancing the Sociological Imagination* (Paradigm Publishers) and, coedited with Sing C. Chew, *Structure, Culture, and History: Recent Issues in Social Theory* (Rowman & Littlefield). His new book, with Jason S. Ulsperger, *Elder Care Catastrophe: Rituals of Abuse in Nursing Homes—And What You Can Do About It,* is forthcoming with Paradigm Publishers. Excerpts from his work are on the website www.sociological-imagination.org. His email address is david.knottnerus@okstate.edu.